LIFE THROUGH THE AGES II

Life of the Past

James O. Farlow, editor

LIFE THROUGH THE AGES II

Twenty-first Century Visions of Prehistory

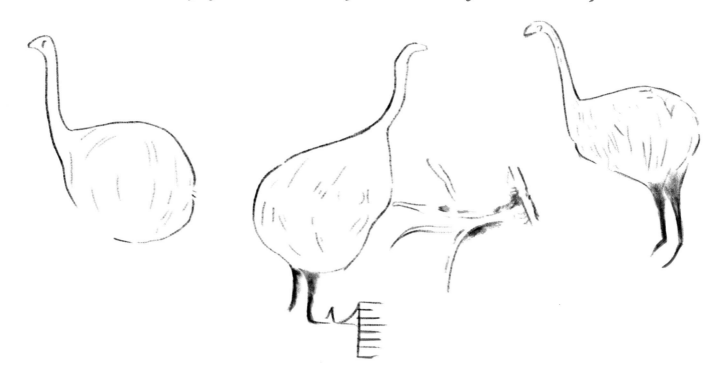

WRITTEN AND ILLUSTRATED BY

MARK P. WITTON

INDIANA UNIVERSITY PRESS

This book is a publication of

Indiana University Press
Office of Scholarly Publishing
Herman B Wells Library 350
1320 East 10th Street
Bloomington, Indiana 47405 USA

iupress.org

Manufactured in Canada

Second printing 2021

Library of Congress Cataloging-in-Publication Data

Names: Witton, Mark P., 1984- author, illustrator. | Knight, Charles Robert, 1874-1953. Life through the ages.
Title: Life through the ages II : twenty-first century visions of prehistory / written and illustrated by Mark P. Witton.
Description: Bloomington, Indiana : Indiana University Press, [2020] | Series: Life of the past | Includes bibliographical references.
Identifiers: LCCN 2019049819 (print) | LCCN 2019049820 (ebook) | ISBN 9780253048110 (hardback) | ISBN 9780253048141 (ebook)
Subjects: LCSH: Paleontology.
Classification: LCC QE763 .W58 2020 (print) | LCC QE763 (ebook) | DDC 560—dc23
LC record available at https://lccn.loc.gov/2019049819
LC ebook record available at https://lccn.loc.gov/2019049820

Contents

Acknowledgments

ALL OF US INVOLVED IN SCIENCE STAND ON THE SHOULDERS OF the generations of researchers who came before us, and books such as this one represent the summation of work from hundreds, maybe thousands, of individuals. Many thanks are owed to these folks who help us uncover the history of life and our planet, as well as those who work—often in trying circumstances—to understand and rectify our current biodiversity crisis. It's hard to write a book about the evolution of life without feeling more invested in the state of life on Earth and the future of this planet. I hope reading this book will thus inspire similar thoughts in others.

Some specific individuals must be acknowledged for their contributions to this book and their encouragement about its content. This work has benefited from input by Victoria Arbour, Nathan Barling, James Boyle, Markus Bühler, Richard Butler, Vicky Coules, Gary Dunham, Jim Farlow, Mike Habib, Luke Hauser, David Hone, Christian Kammerer, Julian Kiely, Darren Naish, the research staff at National Museums Scotland, Felipe Pinheiro, Steve Sweetman, Mike Taylor, and Mathew Wedel. There are possibly others whom I have forgotten to mention: if you're among the omitted, feel free to demand that I buy you a beverage of your choosing next time we meet.

My ability to write educational books and create art is supported by a number of patrons who supply me with a monthly salary at Patreon. com. You guys have made, and continue to make, a huge contribution to my life, for which I'm sincerely thankful. I hope this book justifies your very kind support of my work.

My parents, Paul and Carol Witton, need a mention for their continued patience with a son whom they see increasingly rarely, owing to my being ever busier with different projects that bleed into vacation time and weekends. (I promise I'm not just putting you on speaker phone while I work through your phone calls, honest.) But please spare most thought for four-time book widow Georgia Witton-Maclean, who somehow still puts up with my long, late work hours and my continuous gibbering about whatever cool thing I've been painting or writing about, and demands only that I watch *Deep Space Nine* with her in return. She even read this entire book prior to publication to help check for errors and typos. She's quite OK, that wife of mine. But that's our little secret— don't tell her I said that.

LIFE THROUGH THE AGES II

Introduction

In the Shadow of Knight

THE STORY OF LIFE ON EARTH IS HARDLY A NEW TOPIC FOR AU-thors and illustrators. Popular books on this subject have existed since Franz Unger's 1851 *Die Urwelt in ihren verschiedenen Bildungsperioden* (*The primitive world in its different periods of formation*), a landmark work that described and illustrated (courtesy of artist Josef Juwasseg) the changing environments and inhabitants of our planet for the very first time. Countless examples of the same concept have appeared since then, created by authors and illustrators of varying backgrounds and levels of expertise. Most have largely been forgotten, but the 1946 book *Life through the Ages* is fondly remembered and, thanks to modern commemorative editions, it remains in print well over 70 years since its first publication. The ongoing popularity of *Life through the Ages* has almost certainly been helped by the fact that its author and illustrator is one of the most celebrated and influential artists of extinct animals to have ever lived: Charles Robert Knight (1874–1953).

Nowadays, we consider Knight a *paleoartist*: an individual who restores the life appearance of fossil animals and ancient environments using paleontological and geological data, supplemented by a firm understanding of modern natural history to fill gaps in our knowledge of prehistoric worlds. Although Knight's career saw him capture many natural history subjects, he is probably most famous and fondly remembered for his depictions of prehistory. The discipline of paleoart is as old as paleontological science, stretching back to at least the year 1800, and we can view Knight's professional life, which ran from the 1890s to the early 1950s, as bridging the nineteenth-century foundation of paleoart with a more modern, established period characterizing the mid-twentieth century. Much of the contrast between these eras reflects the rapid accumulation of paleontological knowledge that occurred in the late nineteenth century. Paleoartists working in the early 1800s often

had only scrappy fossils to work from, resulting in reconstructions that, though sometimes surprisingly insightful considering the material they were based on, were not close approximations of their subject species. The discovery of superior fossils in the latter half of the 1800s allowed for new reconstructions that eclipsed the scientific merit of their predecessors. For dinosaurs, in particular, many of these discoveries were being made in the western United States by museum teams from the northeast of the country. As a young and talented natural history artist situated around New York City in the 1890s, Knight was in a prime position to capitalize on this surge of interest in American fossils. In 1894, his habit of sketching animals and specimens in the American Museum of Natural History (AMNH) was recognized by museum staff, and he was asked to restore the life appearance of the extinct piglike mammal *Elotherium* (now *Entelodon*). Knight's career as a paleoartist was thus launched, and thereafter he spent much of his professional life recreating extinct animals in various artforms.

For two decades Knight worked closely with the director of the AMNH, Henry Fairfield Osborn, who promoted Knight's work heavily. Osborn seems to have regarded Knight as a museum "brand" and pushed his work both to advertise the museum and to spread AMNH influence to other institutions. Knight's work became such a beloved component of the museum's exhibits that installations were eventually designed with his artwork in mind: it was important for fossil specimens to be associated with, but not to obscure, his murals and illustrations. Still, Osborn also saw Knight primarily as an artist, not an independent scientific intellect. He referred to Knight's AMNH works as "Osborn–Knight restorations," and in some instances he used Knight's work to visualize his idiosyncratic and infamous ideas on human evolution. Though they developed a productive and successful partnership, Knight and Osborn did not always work in harmony; the two men often disagreed on matters of artistry, science, and artistic ownership. Their working relationship came to an end in 1928, when Knight agreed to a commission from the Field Museum in Chicago. For Knight, distancing himself from Osborn was probably to his benefit, as it demonstrated that he could produce excellent paleoartworks without Osborn's direction. Osborn, however, thought Knight would flounder without his support and was critical of his later work, including his iconic murals for the Field Museum and the Los Angeles County Museum. Despite their less-than-amicable professional split, Osborn and Knight remained friends until Osborn died in 1935, with Knight writing fondly of his colleague after his passing.

Knight's fame and reputation among scholars meant his work became a stamp of quality for any paleontological product from the late nineteenth and early twentieth centuries, almost as if he were somehow the official, licensed artist of extinct life. Knight was just one of many paleoartists working in the late nineteenth and early twentieth centuries, but he dominates histories of the discipline at this time. Indeed, his reputation is so grand that it often eclipses the nineteenth-century artists who created and shaped paleoart in the first place. To some people, Knight *is* the early history of paleoart, or at least the only history worth knowing. He is unprecedented among early paleoartists for his posthumous documentation, making him more than just a name associated with a few paintings: he is a fleshed-out historical figure. In addition to an abridged, posthumously published autobiography (Knight 2005), his work and life have been celebrated through collections of his artwork, correspondence, and biographical accounts (e.g., Czerkas and Glut 1982; Paul 1996; Stout 2002; Berman 2003; Milner 2012; Lescaze and Ford 2017). He is also a recurring character in historical accounts of American science (e.g., Davidson 2008; Clark 2010; Sommer 2016), as well as biographies about influential figures such as Osborn (e.g., Regal 2002). He is, by far, the best-documented paleoartist of all time, and other individuals just as important to early paleoartistry as Knight—such as Benjamin Waterhouse Hawkins, Josef Juwasseg, Edouard Riou, and Zdeněk Burian—are dwarfed by the continued attention and accolade that he receives. (I stress that this comment is not an argument for less attention on Knight, but a reflection on the need for increased scholarly interest in the history of paleoart in general. Knight's documentation is relatively modest in the grand scheme of historical figures, and he is

exceptional primarily because other paleoartists receive so little popular and academic interest.)

Knight's paleoartworks have a high level of technical proficiency and obvious influence from his skills as a traditional natural history and animal artist. His ability to composite exotic extinct species into believable landscapes makes his work as enjoyable today as it was a century ago, even if some of his depictions have become dated. The fact that his work remains on display in Chicago, New York, and Los Angeles long after the work of other artists has been replaced by newer, more contemporary works is testament to his talent and vision. His artwork is all the more remarkable because his eyesight was extremely poor, the combined result of a severe astigmatism and a childhood accident. He was considered legally blind, and it was only through the aid of special glasses that he was able to see. His famous AMNH murals were completed with the aid of assistants, reproducing paintings Knight executed at much smaller size.

Knight is generally not remembered for what seem to be his personally favored subjects or scientifically best work. He is mostly discussed in the context of his dinosaur art, but his real affinity and sharpest scientific insight was for mammals. His writings make no effort to hide his enjoyment of capturing mammals in art, especially elephants, cats, and early humans, whereas he often belittles nonmammalian subjects. In *Life through the Ages*, Knight describes an Asian elephant as "magnificent" and (referring to one individual he knew well) "thoroughly lady-like" (1946, 36), while *Stegosaurus* is "the stupidest member of a very moronic family" (14) (presumptions about animal intelligence are common in Knight's writings, and are often so glib as to be humorous, whether intentionally or not). His other books show similar biases. *Before the Dawn of History* (1935) leans toward mammal evolution and human prehistory in both text and illustrative coverage, and his 1947 *Animal Anatomy and Psychology for Artists and Laymen* (titled since 1959 as *Animal Drawing: Anatomy and Action for Artists*) devotes 82 pages to mammal anatomy, poses, and behavior, and just 14 pages to birds, reptiles, and invertebrates. His final book—*Prehistoric Man: The Great Adventurer* (1949)—is a 330-page summary of hominid evolution, showing his keen academic interest in human origins. This was his most scholarly tome, featuring very little artwork alongside the extensive text. Further evidence of Knight's zeal for mammalian subjects is found in his autobiography (Knight 2005), where he remarks favorably about

Knight's *Brontosaurus* illustration from *Life through the Ages* (1946). Knight is perhaps best-known for his dinosaur art, though dinosaurs do not seem to have been his favorite art subjects, nor are they—scientifically speaking—his best work.

drawing mammals at zoos and describes his travels to Europe to see hominid fossil localities.

The combination of Knight's talent and keen interest in mammals explains his terrific artistic and scientific successes with portraying these creatures. His scenes of fossil reptiles however, while just as well composed, are anatomically peculiar to modern eyes. The detailed diagrams of mammalian anatomy showcased in his 1947 *Animal Anatomy for Artists and Laymen* leave no doubt that Knight knew how animal bodies were put together—how muscle shape and size are determined by skeletal landmarks, where shapes of the face and body conform to underlying bones, and so on—but his nonmammal works sometimes translate skeletal anatomy into restored forms only loosely. For instance, his dinosaurs have thighs that are far too slim relative to the massive pelvic bones they were attached to; have proportions that are peculiar even when compared to fossils known at that time; and have faces that sometimes deviate from their skull contours, particularly in carnivorous species. His reptiles also often lack the dynamism and nuance evident in his mammal art, mostly having relatively static poses and only rarely showing complex behavior, such as parenting or herding, despite these being commonplace in his prehistoric mammals.

What makes this discrepancy fascinating to scholars of paleoart is its indication of cultural attitudes overriding an otherwise well-honed scientific eye. Though continued publication has given Knight's work an ageless quality, like any scientific artwork it was informed by the ideologies and theory of its time. It's worth highlighting the context Knight was creating his artwork in, and how he—even working with leading scientists and advisers—would have understood the prehistoric world. Many ideas and concepts about extinct life that we now take as established fact were uncertain, or even entirely unimagined, to people of Knight's day and age. For example, Knight would never have had a firm idea of how old his paleoart subjects were, the age of Earth and its geological periods being poorly constrained until the mid-1940s. He would have known continental drift only as a controversial theory favored by a few geologists. He died the same year that DNA was discovered, and thus missed out on many fundamental revelations about evolutionary processes. He also shared the erroneous view of evolution as a continued optimization of nature toward the modern day, where older creatures were inferior to newer ones. Mammals, for example, must be smarter, more behaviorally complex, and physically superior to

dinosaurs because they outlasted them. In *Before the Dawn of History*, Knight describes mammals as "leaders among the created things" (1935, 8) while dinosaurs are "weird, monstrous and bizarre" (7). Knight (with a small handful of famous artistic exceptions) thus followed the twentieth-century idea that dinosaurs were slow, sluggish creatures ill-suited to any strenuous activity or complex behavior, despite osteological evidence to the contrary. We can only wonder how Knight's extinct reptiles might have looked if he had been as anatomically objective in their reconstruction as he was with his mammals, and if his attitudes toward nonmammalian subjects had been a little more forward-thinking. Sadly, Knight seems to have never written about dinosaur anatomy in much detail and we can only speculate on how he rationalized his reconstructions. He certainly had strong opinions on their portrayal, however. Records of his correspondence include a 1937 letter to a newspaper about Emmet Sullivan's highly stylized dinosaur sculptures in a park in Rapid City, South Dakota, in which he berated the sculptures as "amateur and foolish" (Milner 2012, 148). We have to wonder how he viewed the work of such contemporary artists as Gerhard Heilmann (1859–1946) and Harry Govier Seeley (1839–1909), who envisaged dinosaurs and flying reptiles in more anatomically correct and progressive ways. Ultimately, Knight's talent, prestige, and association with top academic institutions gave his reptile reconstructions greater cultural weight than was given to scientifically superior contemporary work, and he went on to have a huge influence on popular culture.

The high quality and continued use of Knight's work has seen him become one of the most copied and referenced paleoartists of all time. Perhaps only the Czech artist Zdeněk Burian (1905–1981), in many respects Knight's successor in paleoart mastery, can claim similar treatment. Countless artists have used Knight's portfolio as inspiration for original works or produced thinly disguised reworkings of his compositions. Decades of replicating Knight's takes on prehistoric life place him as the source for several long-standing paleoart tropes and clichés. Some Knightian conventions—such as the establishment of *Triceratops* and *Tyrannosaurus* as enemies set to battle each other across generations of artwork (the first portrait of *Tyrannosaurus*, which Knight drew in 1905, featured this animal lurking close to a group of *Triceratops*; he revisited this theme more dramatically in his 1930 Field Museum mural)—are understandably artistically appealing for their heroic, almost legendary quality, but other Knightian memes are far

A modern take on a sauropod dinosaur, *Diplodocus carnegii*, to compare with Knight's 1946 example. Note the powerful musculature around the top of the hindlimb and tail base, a reality of dinosaur anatomy that Knight mysteriously overlooked despite his expertise in restoring extinct animals. Some aspects of Knight's dinosaurs reflect the culture of his time more than the objective nature of their anatomy.

more idiosyncratic. Examples include his bird-chasing *Ornitholestes* and stooping *Allosaurus* scavenging a dinosaur tail, both of which have been replicated over and over by artists despite their specificity and, in some cases, problematic science.

But it was not only illustrators who felt Knight's influence. Early filmmakers also referenced his paintings, most famously to create prehistoric animals for *The Lost World* (1925), *King Kong* (1933), and the *Rite of Spring* sequence in Disney's *Fantasia* (1940). Over a decade after his death, *One Million Years BC* (1966) and *Valley of Gwangi* (1969) still used his work as a reference for their prehistoric creatures. While most of the filmmakers behind these projects openly acknowledge Knight as their source (e.g., Harryhausen and Dalton 2003; Harryhausen's foreword in Knight 2005), his name is entirely absent from anything to do with *Fantasia*, a curious circumstance given the film's prehistoric animal sequence featuring numerous Knightian reconstructions and callbacks—far more than would be expected by chance (Davidson 2008). This may not reflect a simple lack of documentation about the making of Disney's animation, either. A short article about the production of the *Rite of Spring* sequence, published in 1941 by the AMNH paleontologist who consulted on the film, Barnum Brown, mentions passing AMNH-endorsed prehistoric animal restorations to Disney as a basis for their creature designs. Given the historic context, Brown was surely supplying Disney with Knight's work, but the artist curiously gets no mention. Was this an oversight, or had Knight's departure from the AMNH diminished their desire to promote him as part of their "brand"? In any case, Knight has been more readily namechecked in fiction, such as in Ray Bradbury's 1983 story *Besides a Dinosaur, Whatta Ya Wanna Be When You Grow Up?* Knight is mentioned twice in this short story, once as "a poet with a brush . . . Shakespeare on a wall" and later as "the man who sees through time, and *paints* it!" Bradbury would go on to write a short forward to Knight's autobiography (Knight 2005).

It was only toward the end of the twentieth century that Knight's influence began to wane, perhaps because new fossil discoveries and theories were modifying our concepts of some prehistoric animals beyond the point where his depictions were usable, because new cultural touchstones (such as the 1993 film *Jurassic Park*) redefined expectations for paleoart, and because novel paleoartistic methods were being established by a new generation of artists. But his legacy remained as strong as ever, as evidenced by the continued publication of books and articles about his work. Even today, it's a challenge for professional paleoartists to be seen outside of the shadow cast by one of the giants of the discipline, and perhaps justifiably so.

A *LIFE THROUGH THE AGES* SEQUEL
SEVEN DECADES ON

Life through the Ages is perhaps the best known and most accessible of Knight's books. Its structure—a series of plates, each mirrored by a single page of descriptive text—is similar to the earlier *Before the Dawn of History*, but it lacks the expanded introduction, has fewer plates, and contains illustrations of living species to place his fossil subjects in context. It also omits the ordering of subjects by geological age that was used in *Before the Dawn of History*, which in addition compiles many of Knight's famous murals. The art of *Life through the Ages* instead comprises original charcoal sketches and a few Knight artworks borrowed from museum libraries and magazines. It remains highly readable even today and is a terrific way to sample our understanding of animal life, both prehistoric and modern, from the early twentieth century. As in all of Knight's written work, his text is confident, lively, and charismatic, perhaps even more so to modern readers armed with seven decades of scientific hindsight.

Life through the Ages has been continually reprinted since 1946 (most recently in a commemorative edition by Indiana University Press [Knight 2001]), making it a consistently available opportunity to appreciate early twentieth-century views on animal biology and evolution. It's this quality that makes *Life through the Ages* a perfect book to write a sequel to. It would have been easy to produce another book outlining the marvels of evolution under a different title, but writing a companion piece to *Life through the Ages* allows for direct comparison of modern ideas with those of several decades ago. I hope readers of this book will track down a copy of the original for this purpose. Both books are designed for general audiences and are illustration-heavy, so this exercise can be performed by readers of any age. Bear in mind that the artistic differences between these books reflect not just the whims of their authors, but also several generations of scientific discovery and research, as well as the distinct cultures of the mid-1940s and late 2010s. Some content of this book is specifically referential to the original *Life through the Ages* to aid this comparison, showing Knight's subjects revisited as we understand them today. In several cases, Knight's original images provided compositional foundations for these new artworks, while others contain subtler homages to Knightian conventions and tone. Where our understanding has moved on too far, the new illustrations share subject matter only. But when looking at these modern visions of prehistory, remember that they too will become dated. All books on science, and the art they contain, are products of their time—mere snapshots of understanding. Perhaps in another seventy to eighty years, another entry in the *Life through the Ages* canon will give a new generation a chance to see how much further our understanding of prehistory has moved on.

As in the original book, this sequel explores ancient and modern life in the tradition of illustrations and descriptive text, foregoing diagrams and infographics in favor of life and landscape reconstructions. I have taken the opportunity not only to replicate Knight's format but also to enhance it, using full-color images and sixty-two rather than thirty-three illustrations in the main portion of the book. Our discussions of the paintings can also be more factual and detailed, thanks not only to our improved knowledge of fossil life but also to the greater availability of scientific literature to modern authors. In the interest of transparency about the scientific content of the book, I have added an appendix with notes about the paleontological data used in each image. These are comments that would derail our narrative if they were included in the main text, but they serve to establish which aspects of the restorations are based on facts, which are informed guesswork, and which are pure speculation. All paleoart has a blend of these components, but without notes from the creators of the images, we are often left to figure out their proportions ourselves.

An attempt has also been made to balance this sequel fairly to different geological divisions. It would be impossible to cover the evolution of life entirely in a book of any practical size, but we can at least recognize that the story of life is not simply the story of humanity. The evolution of land animals did not mark the cessation of fish evolution, just as the rise of mammals did not prohibit innovations in plant and insect lineages. We must also recognize the role that extinction has had in shaping life on Earth; to that end, several of the most dramatic extinction events in history are the focus of four paintings. Some grounding in the theories behind geological time, the relationships of organisms, and the paleoartistic process has also been added as a primer for readers who are less familiar with these concepts, as well as to establish how the disciplines of geology, evolutionary biology, and paleoart have moved on since 1946. I have given less emphasis to living species than Knight did, because, in our modern age, popular documentation of living forms has increased markedly over that of fossil animals. The 1940s was a different age not only for photography but also for photographic reproduction in books. We now enjoy spectacular images of animals and plants wherever we look, as well as greater availability of facts about these species. Many fossil animals, in contrast, remain poorly represented in art, and their paleobiology is rarely discussed outside of academic papers. I thus feel that readers will benefit more from their coverage here than from greater documentation of living species.

A final, and perhaps most important, distinction in this sequel is a subtle change of tone. The early and mid-twentieth century had a vastly different view of human development, population, and our relationship with wildlife and natural environments. Growing realization about our shrinking, weakening biosphere necessitates a greater reverence for the natural world than was generally held in the 1940s, and this underscores the urgent need for its preservation. More than ever, we need wider understanding of concepts like ecology, extinction, biodiversity, the reality of geological time, and the way that living beings influence global habitats. We must appreciate that populations of organisms—whether they are the first plants on land, herds of megaherbivores converting leaves into greenhouse gases, or primates driving motorcars—can cause planetary changes that affect life for millions of years, a scale of time that we cannot easily comprehend. Our current concept of humanity's impact on Earth is dangerously short-sighted compared to the vastness and complexities of planetary ecology and geological time, and we must respond to overwhelming data indicating that we stand on the brink of environmental and climatic catastrophe. There has never been a more important time for understanding our place in the natural world, the evolutionary history that we are part of, and the way that organisms—including ourselves—shape the future of planet Earth.

The Antiquity of Earth, and Its Geological Divisions

THE AGE OF EARTH HAS FASCINATED SCHOLARS FOR MILLENNIA, but for most of civilized history our best guesses of its age were rough calculations based on historical records or biblical texts. Most of these early estimates were relatively low: a planet Earth just a few thousand years old. It was not until scientific processes started to dominate scholarly thinking in the seventeenth and eighteenth centuries that we began to fathom the true enormity of Deep Time. Pioneering geologists were unable to provide a specific figure for the age of Earth, but their appreciation for how long it took sedimentary rocks to form indicated an Earth that was far, far older than anyone had previously anticipated. By the late nineteenth century, estimates of the time needed to accumulate modern rock layers and seawater salinity pointed to an age of at least one hundred million years. But our best estimate, developed during pioneering work on radioactive decay rates in the 1950s, raised the predicted age much further: 4.55 *billion* years. This figure has since stood as a robust calculation of Earth's age, correlating with our geological record (the oldest rocks on Earth are about 4.4 billion years old) and estimates of the age of our solar system (4.6 billion years old).

Many of Earth's rocks are deposited in layers, stacked atop one another in order of age: the oldest at the bottom, and the most recent at the top. Because the properties of rocks are highly characteristic of the conditions in which they were created, this layering system allows us not only to date rocks based on their relationships to one another but also to track changes in environment and climate through time. This provides temporal and environmental context for the fossils contained in each rock layer and permits reconstruction of changes in Earth's biosphere throughout history. We use widespread characteristic rock horizons or fossils to correlate geological events (and thus time) in outcrops found far away from one another, and we can precisely date some rock layers using radiometric dating techniques. These measure the amount of decayed atomic matter in certain minerals against their undecayed variants, allowing us to compute their age from the known decay rates of those elements. Collectively, these dating and correlating techniques have led to the creation of the geological timescale, a chronology of Earth's history as told by the rock record. The timescale continues to be updated with new dates and divisions as new data becomes available, although modern revisions generally represent fine-tuning of a robust and mature scientific model rather than textbook-shattering changes to the entire system.

Many readers will be familiar with the fact that geological time is divided into categories: terms like *Jurassic* and *Pleistocene* are familiar examples of the names given to geological eons, eras, periods, epochs, and so on. These divisions are not arbitrarily assigned to ancient dates but are contingent on widely mapped geological or biological events, such as a major turnover in the biosphere (perhaps reflecting a mass extinction), the appearance of a specific type of fossil, or a geographically widespread and characteristic rock layer. The result is a system that reflects major developments in the history of the planet, and one that provides increasingly precise characterization of environmental, climatic, and biological conditions as we move from broad to fine divisions of time.

As is evident from the geological timescale opposite, our ability to tease apart significant periods of Earth's history improves in younger rocks. This reflects the fact that older rocks are rarer—having had longer to be buried, deformed, or destroyed by geological action and erosion—and the related improved quality of the fossil record in younger strata.

Geological Time

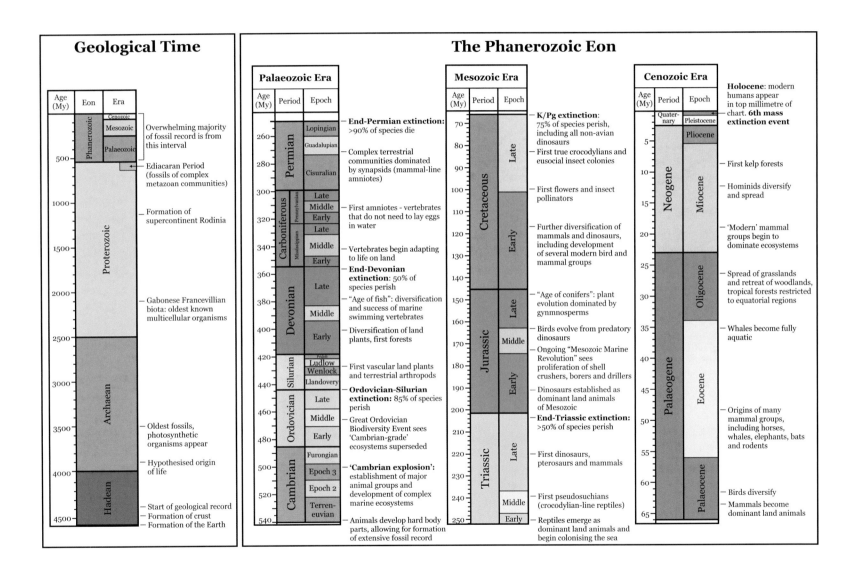

Age (My)	Eon	Era
	Phanerozoic	Cenozoic
		Mesozoic
500		Palaeozoic
1000	Proterozoic	
1500		
2000		
2500		
3000	Archaean	
3500		
4000		
4500	Hadean	

- Overwhelming majority of fossil record is from this interval
- Ediacaran Period (fossils of complex metazoan communities)
- Formation of supercontinent Rodinia
- Gabonese Francevillian biota: oldest known multicellular organisms
- Oldest fossils, photosynthetic organisms appear
- Hypothesised origin of life
- Start of geological record
- Formation of crust
- Formation of the Earth

The Phanerozoic Eon

Palaeozoic Era

Age (My)	Period	Epoch
260	Permian	Lopingian
		Guadalupian
280		Cisuralian
300	Carboniferous (Pennsylvanian)	Late
320		Middle
		Early
340	Carboniferous (Mississippian)	Late
		Middle
		Early
360	Devonian	Late
380		Middle
400		Early
420	Silurian	Pridoli / Ludlow
		Wenlock
440		Llandovery
460	Ordovician	Late
		Middle
480		Early
500	Cambrian	Furongian
		Epoch 3
520		Epoch 2
540		Terreneuvian

- **End-Permian extinction:** >90% of species die
- Complex terrestrial communities dominated by synapsids (mammal-line amniotes)
- First amniotes - vertebrates that do not need to lay eggs in water
- Vertebrates begin adapting to life on land
- **End-Devonian extinction:** 50% of species perish
- "Age of fish": diversification and success of marine swimming vertebrates
- Diversification of land plants, first forests
- First vascular land plants and terrestrial arthropods
- **Ordovician-Silurian extinction:** 85% of species perish
- Great Ordovician Biodiversity Event sees 'Cambrian-grade' ecosystems superseded
- **'Cambrian explosion':** establishment of major animal groups and development of complex marine ecosystems
- Animals develop hard body parts, allowing for formation of extensive fossil record

Mesozoic Era

Age (My)	Period	Epoch
70	Cretaceous	Late
80		
90		
100		Early
110		
120		
130		
140		
150	Jurassic	Late
160		
170		Middle
180		Early
190		
200		
210	Triassic	Late
220		
230		
240		Middle
250		Early

- **K/Pg extinction:** 75% of species perish, including all non-avian dinosaurs
- First true crocodylians and eusocial insect colonies
- First flowers and insect pollinators
- Further diversification of mammals and dinosaurs, including development of several modern bird and mammal groups
- "Age of conifers": plant evolution dominated by gynmnosperms
- Birds evolve from predatory dinosaurs
- Ongoing "Mesozoic Marine Revolution" sees proliferation of shell crushers, borers and drillers
- Dinosaurs established as dominant land animals of Mesozoic
- **End-Triassic extinction:** >50% of species perish
- First dinosaurs, pterosaurs and mammals
- First pseudosuchians (crocodylian-line reptiles)
- Reptiles emerge as dominant land animals and begin colonising the sea

Cenozoic Era

Age (My)	Period	Epoch
	Quaternary	Pleistocene
5	Neogene	Pliocene
10		Miocene
15		
20		
25	Palaeogene	Oligocene
30		
35		
40		Eocene
45		
50		
55		
60		Palaeocene
65		

Holocene: modern humans appear in top millimetre of chart. **6th mass extinction event**

- First kelp forests
- Hominids diversify and spread
- 'Modern' mammal groups begin to dominate ecosystems
- Spread of grasslands and retreat of woodlands, tropical forests restricted to equatorial regions
- Whales become fully aquatic
- Origins of many mammal groups, including horses, whales, elephants, bats and rodents
- Birds diversify
- Mammals become dominant land animals

11

The Relationships of Earth's Inhabitants

EVERYTHING LIVING ON OUR PLANET IS A DESCENDENT OF Earth's first microbial lifeforms, linked to them through an unbroken chain of genetic material passed down through billions of years by reproducing organisms. It's sobering to consider this long, unifying link among all life on Earth—which, of course, includes you and me. We tend to limit our gaze to our immediate ape ancestors when pondering our relationships with other species, but this is just the start of our connectedness to other life. Go back further and we share ancestry with countless other species: other mammals, egg-laying land vertebrates, fish, marine organisms like sea squirts and echinoderms, early animals like corals and sponges, and then on to fungi, plants, and many different types of microbes. We are shoots on an evolutionary tree that has, over the course of Earth's history, grown billions of branches from roots buried at the very origin of our planet.

The tremendous biodiversity of our planet is owed to two phenomena: (1) the reproduction and recombination of RNA and DNA (ribonucleic acid and deoxyribonucleic acid, respectively, the molecules that make up the genes of different organisms) and (2) the optimizing process of natural selection. RNA and DNA are the molecular codes that tell organisms how to build cells and tissues, and any changes to them (deliberately, such as combining DNA of two individuals in sexual reproduction, or accidentally, through gene replication errors—otherwise known as mutations) alters how our anatomy is created or functions. These changes mostly have no significant effect, but they sometimes influence an organism's ability to survive in its respective environment. Individuals that receive an advantageous change stand a better chance of reproducing and thus passing on their genes, while those that inherit a deleterious genetic alteration have lessened reproductive probability. This is "natural selection" in action. As organisms spread into different habitats or environmental conditions change over time, either natural selection modifies organisms into species that are well suited for the conditions they live in, or their lineage struggles to adapt and ultimately perishes. Because organisms must be well-rounded entities, capable of performing many actions (e.g., obtaining food, avoiding predation, protecting our tissues against harmful environmental conditions), they can never be "perfected" to one task. Our bodies are adaptive compromises, proficient enough at everything we need to do to survive and pass on our genetic material, but are never so specialized to one task that other crucial functions are impeded.

Evolutionary relationships between organisms are revealed through shared anatomical and developmental characteristics. Greater similarity indicates a relatively close ancestry, and fewer shared features indicate a more distant relationship. Humans, for example, are so anatomically and genetically like chimpanzees and gorillas that our close and geologically recent ancestry is in no doubt. But our similarity to plants is expressed in only some aspects of our cellular anatomy and genetic makeup, indicating that our respective evolutionary branches have been developing separately for a very long time. We can use genetic data to determine evolutionary relationships among living species, but for fossil organisms we largely rely on anatomical features—the shapes and characteristics of bones, shells, and other tissues—for this purpose. Our record of the history of life is imperfect and we do not understand where all living organisms belong in our evolutionary tree, but the general picture of life's evolution, and the relationships of species on many specific branches, are increasingly well understood.

As our grip on evolutionary theory has developed, it has become apparent that traditional schemes for classifying life—ranking it within categories like *kingdoms*, *orders*, and *families*—are misleading and arbitrary. Many biologists and paleontologists now refer to groups or *clades* of animals, collectives of species united by shared features without a set "rank" in the grand order of life. This reflects the true continuum of evolution better than assigning arbitrary significance to certain groups of species, and it is the approach used throughout this book.

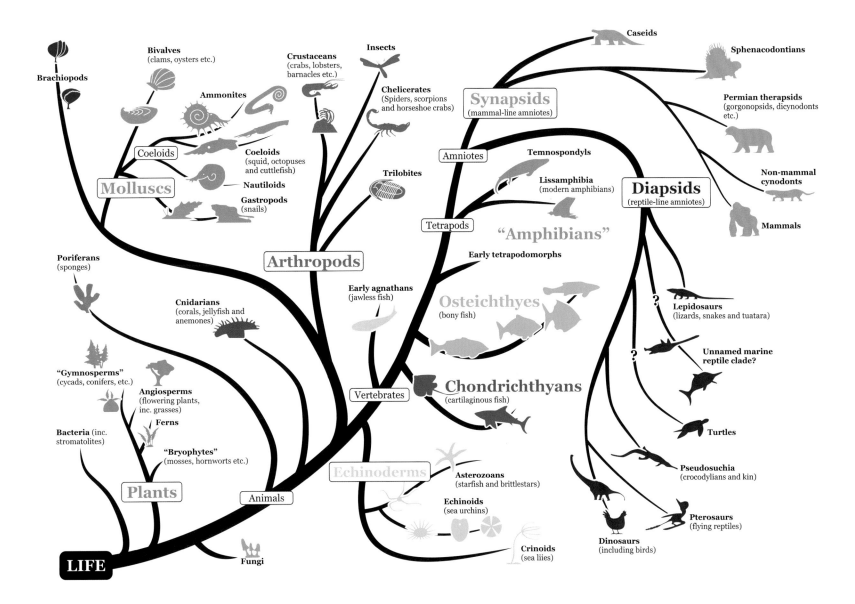

13

How Prehistoric Animals Are Reconstructed in Art

THE PALEOART PROCESS IS NOT WELL KNOWN OUTSIDE OF A sphere of individuals with a keen interest in fossil animals, so it is worth outlining here to put the following illustrations in context. I stress the use of the word *process* in that sentence: paleoartworks are not just loosely inspired by fossils, but are the result of research into the anatomy, evolutionary relationships, and geological context of an extinct subject species. Well-researched paleoart is a visual hypothesis of the life appearance of a fossil organism and its environment. The incompleteness of the fossil record means that speculation and reasoned prediction have a role in this process, and this prevents paleoart from being a fully scientific discipline, but the percentage of unknown data varies from species to species. Some extinct organisms are known in enough detail that our reconstructions are probably reasonable approximates of their actual appearance (even aspects of color can now be predicted for animals that lived over one hundred million years ago, so long as we have sufficiently high-quality fossils), while others can be sketched only vaguely from scant remains. It can be hard to ascertain the reliability of a reconstruction at face value, and for this reason the appendix at the back of this book clarifies certain decisions and data used in each of these paintings.

The basic steps of reconstructing a fossil animal are as Knight described them in his 1935 book *Before the Dawn of History*, but palaeoartistry has become much more technical since his time. In recent decades, paleontological science has adopted increasingly rigorous scientific practices to produce more robust and detailed understandings of evolution, the history of our planet, and the paleobiology of fossil species. Modern paleoart has to reflect these advances to create scientifically credible artworks. We rarely know everything about the life appearance of extinct animals, but we increasingly know what *not* to portray in our artworks, and this narrowing of our reconstruction options moves our visual hypotheses in more credible directions.

Beyond information gathering, the first step in many paleoartworks is a skeletal reconstruction: a process where the artist arranges fossil bones in a lifelike pose to fully understand the basic proportions of a fossil species. Where bones are missing or damaged, information from other individuals of that species or a close relative are used to fill the gaps. Information from fossil trackways can help determine aspects of pose and gait. With an understanding of the skeleton in place, the artist can add musculature over the top. This is performed in the anatomical context of close living species, transferring landmarks of muscle position to the extinct animal to create a defensible arrangement of muscles across the entire skeleton. Fatty tissues are virtually unrepresented in the fossil record, but an artist may speculatively add them at this stage, typically by considering where closely related living animals store their fat, or where fat storage would be effectively placed in the subject animal (e.g., would it be localized somewhere to minimize inconvenience, or would it be distributed across the body as insulation?). Details of skin—anatomies like scales, feathers, and fur—are either based on direct fossil evidence (features like scales and feathers are rare in fossils, but are preserved more often than might be intuitively assumed) or predicted using close relatives (even living species, in some cases). There is a lot of uncertainty about the skin types of certain extinct clades because, as plainly demonstrated by modern species, even closely related animals can differ markedly in attributes like fur or feather length, scale size, and so on. Colors and patterning can be determined (partly or completely) for fossil species in a variety of ways, but they are unknown for over 99 percent of extinct animals. Artists therefore mostly create color schemes that suit the predicted lifestyle and habitat of the subject animal. This process completes a model of the appearance of a living animal, but to give it a home among an appropriate landscape, climate, and flora, we must further consult geological and paleobotanical data. So begins another round of research and injection of science into the paleoartwork, each bringing its own challenges and surprises to visualizing prehistory.

The fundamentals of paleoartistry

Skeletal reconstruction
Created using bone morphology and proportions of fossil remains, assembled in a biomechanically plausible pose for the subject. Missing elements are based on closely related species and scaled to an appropriate size.

Muscle reconstruction
Muscles are placed on the skeleton in a configuration that reflects the distribution of muscles in closest living relatives, but with a volume that suits the functional predictions of the subject species' lifestyle.

Footprint data
Fossil trackways and footprints tell us much about extinct animal postures, speeds and gaits, all of which inform predictions of life appearance.

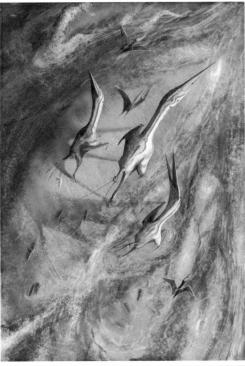

Considering soft-tissue
Fats, skin and related structures are added in line with fossil data (soft-tissues rarely fossilize, but can in exceptional circumstances. Certain facial skin details can be predicted from bone surface textures). Ideally these data are taken from fossils of the subject species but data from closely related, similarly-adapted species can also be used when this is unobtainable. If soft-tissues are completely unknown, likely tissue types are predicted based on the relationships of the subject to other species, and soft-tissue adaptions of ecologically similar living species.

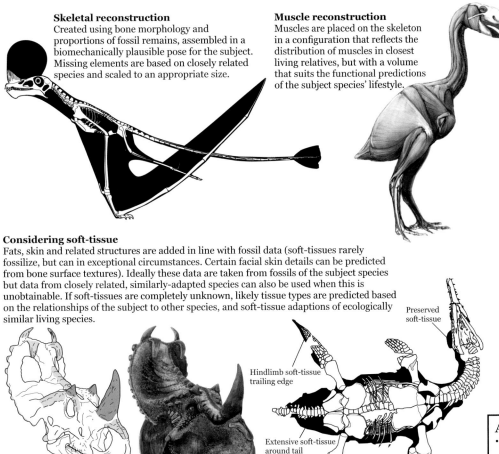

Preserved soft-tissue

Hindlimb soft-tissue trailing edge

Extensive soft-tissue around tail

Forelimb soft-tissue trailing edge

Scale correlates
Cornified sheath ('horn') correlates

Additional critical considerations
- **Geological context of subject species**: informs reconstruction of environment and climate
- **Age and location of subject species:** informs decisions about contemporaneous fauna and flora
- **Palaeobotany:** essential for credible restoration of extinct plant species

The Plates

In Pleistocene northwest Europe, prehistory met modernity as recognisable species of walrus, hyena, and swans existed alongside now extinct southern mammoths, cave lions, woolly rhinoceros, and great auks. The modern natural world emerged gradually from what we class as 'prehistory' without a sudden transformation of flora, fauna or environment, but instead as the product of billions of years of biological evolution, extinction, and environmental change.

Building a Planet Fit for Life (Hadean)

THE STORY OF LIFE ON EARTH BEGINS SEVERAL HUNDRED MILlion years before the first organisms appeared, taking us back 4.55 billion years to the origins of our planet, sun, and solar system. While these events may seem far removed from the origin of life, they were essential in providing the right conditions for life's development: the construction of a global habitat, the gathering of raw materials for organismal bodies, and the positioning of Earth in the right part of the solar system for plentiful, but not excessive, solar light and heat. Though it's probable that earthly conditions are not essential for all life (life-forms on other celestial bodies, if they exist, would have to evolve with biochemistry and environmental tolerances reflecting their own circumstances), Earth's properties and place in the cosmos were key factors in the origin and development of life on this planet.

Before Earth was formed, the atoms and molecules that would one day form living beings were indivisible from those that would create the sun, eight planets, and thousands of celestial bodies that make up our solar system. At this point, our constituent elements were mixed in a vast interstellar molecular cloud, the remnants of long-deceased stars. We existed as masses of hydrogen, helium, some heavier elements, and pockets of dust and gas, hanging out together in a small portion of the Milky Way galaxy. Around 4.6 billion years ago, a fragment of our molecular cloud—perhaps prompted by the energy of a local supernova—began to collapse under its own gravity and coalesced into a star: our sun. This pulled additional portions of our cloud into orbit around the new star, forming a broad ring of jostling dust and gases that collided and fused with one another. As coalescing bodies of cosmic dust created ever-larger celestial objects, the sun's orbit became a bustling ring of planetesimals (solid objects measuring a kilometer or more across) and protoplanets (moon-sized objects), the largest growing rapidly as their increasing gravitational mass hoovered up debris from their orbits. Eventually, most of the dust and debris from this protoplanetary disk stabilized into the order of our solar system: a number of smaller stellar bodies and eight planets, one of which was Earth. But our planet was far from habitable at this stage. For at least several tens of millions of years, if not hundreds of millions, Earth remained an angry red ball floating through space: a highly radioactive planet superheated from collisions with other stellar objects and under continued bombardment from asteroids. Earth's first 500-million-year-long eon is appropriately known as the Hadean—"hellish"—Eon.

But despite the chaotic and violent origins of Earth, its position and properties were promising for life. Earth is a relatively heavy planet and has thus been pulled close to the sun, but it is not so close as to be excessively hot, nor so far away as to receive diminished light and heat energy. Our orbit around the sun is within a narrow temperature band that allows for the existence of liquid water at the planet's surface. Once Earth cooled, likely relatively early in the Hadean Eon, liquid water was able to cover 70 percent of the planet. This not only provided a major habitat for life, but also liberated water for use in our biochemistry. Liquid water is a critical chemical component for all organisms on Earth and life on our planet would be radically different, if it existed at all, without it. Our atmosphere was attained and lost several times in our early history thanks to influences of solar and geological events, but it was ultimately instrumental in the creation of oceans and seas. Once Earth was cool enough (about 4.4 billion years ago), the first clouds began raining the oceans into existence on our newly formed planetary crust. Tectonic activity—the creation, destruction, and motion of crustal plates—began in the Hadean and created the first small continents by the end of the eon. Subsequent tectonic activity would add continental material to these early landmasses to form the continental plates we know today. Associated volcanism transferred complex chemicals from within the Earth to its surface, and these were likely instrumental in the formation of the first nonorganic biomolecules—the raw components of life. From a vast stellar cloud of stardust, this 500-million-year process created a planet with a strong potential for originating and sustaining life.

The Origins of Life (Archaean)

DETERMINING HOW, WHEN, AND WHERE LIFE APPEARED ON Earth has preoccupied humanity for centuries. This question has been considered philosophically, as means to understand humanity's place in the universe, as well as religiously, with countless faiths developing creation myths to account for the advent and development of life. But our best, and realistically only, mechanism for understanding the origins of life lies with scientific approaches. While it remains true that scientists have yet to determine exactly how life began on Earth, it would be wrong to say that nothing is understood about the origins of life at all. Quite the opposite is true, and we are narrowing down how this seemingly miraculous event could happen through fundamental chemical and physical processes.

The oldest incontrovertible evidence for life on Earth is 3.5 billion years old, though other evidence indicates an even earlier origin at 3.9 billion or even 4 billion years ago. All three figures imply that life appeared rapidly after the formation of the planet, despite the early Earth being a turbulent place that would be largely inhospitable to the life of today. Earth of the early Archaean Eon was still experiencing vigorous tectonic activity and was pounded by meteors left over from the formation of the solar system. The oceans and atmosphere, although probably cooler than they had been during the Hadean, were of a composition radically different from those of more recent times.

These conditions might seem like the last place where organic beings could arise, but a planet of churning complex molecules and frequent electrical discharges was probably an ideal setting for the chemical reactions needed to catalyze life. Inorganic matter has some self-organizing properties because of inescapable laws of chemistry and physics, so we can be certain that complex chemicals and biomolecules were forming on ancient Earth without guidance. From these came the fundamental components of organismal cells and, with them, the potential for the first basic lifeforms. There are no physical or chemical components of our bodies that are unique to us as living beings, nor is there evidence that life requires a magical "spark" or supernatural ingredient to be created and sustain itself. We are separated from inorganic forms only by specific processes of maintaining, growing, and reproducing our chemical makeup, these likely being defined and refined at a very basic level billions of years ago by chemical and biological evolution.

It is assumed that our inorganic–organic transition occurred in stages. Although true life has never been experimentally created, some of its earliest stages have been artificially generated in conditions predicted for Hadean or Archaean environments. Raw biomolecules—such as the components of cell membranes and amino acids—have been abiotically constructed from mixes of gases, liquids, and electrical charges, and the addition of certain minerals has been shown to enhance self-organization of these molecules into forms that could become "protocells" under the right conditions. It is not at all unreasonable to assume that the full inorganic–organic transition will one day be fully understood and simulated in laboratory conditions.

Inorganic biomolecules are known to occur in a number of extreme environments, such as volcanoes, mid-ocean ridges, and even interstellar space, giving scientists a number of potential sources of organic material and locations where life could have arisen. Among the most likely settings are mid-ocean ridges, where a combination of heat and abundant organic molecules could have created the first simple cells. We might not imagine life originating at the explosive, magma-spewing mid-ocean ridges we are most familiar with, however: hydrothermal fields—vast plains of limestone towers created by uprisings of geothermal water—are rich in hydrocarbons (the basic components of cell membranes) and are better candidates for life's earliest habitat. These dark, deep-water settings form the subject of the adjacent illustration.

Stromatolites (Archaean–Proterozoic)

ALTHOUGH LIFE APPEARED ON EARTH RELATIVELY QUICKLY, IT seems to have remained relatively low-key, even microscopic, for most of Earth's history. The proliferation of macroscopic, multicellular organisms—those we can easily see with the naked eye—is a relatively recent evolutionary event that occurred only 600 million years ago. This is not to say that early life was not conspicuous in other ways, however. Visitors to Earth as far back as 3.7–3.5 billion years ago had a good chance of seeing stromatolites, mounds of sediment bound together by films of cyanobacteria (simple, photosynthetic cells also known as blue-green algae).

Stromatolites have an excellent fossil record that reaches its zenith about 2.5 billion years ago, before declining as other, more complex life appeared at the end of the Proterozoic Eon. They survive today, but largely in places inhospitable to other organisms, such as lagoons with high salinities. Their evolutionary nemeses are grazing invertebrates, such as snails, which devour defenseless cyanobacteria with such ferocity that stromatolites can now grow only in habitats intolerable to these invertebrate herbivores.

As mounds of bacterially bound sediment, stromatolites look unremarkable. They are significant organisms for many reasons, however. Oxygen-breathing organisms such as ourselves owe much to stromatolites for their transformation of our ancient, carbon dioxide–rich atmosphere into a breathable, oxygen-rich one. Without their billions of years of oxygen-deploying photosynthesis, the story of evolution on Earth would be entirely different. Stromatolites also form the majority of our record of early life and have the accolade of being the oldest known fossils. The most-ancient examples are from Australia and Greenland, and these specimens have an increasingly good chance of being the oldest fossils we'll ever know. Early Archaean rocks are very rare on Earth's surface today and most are so heavily distorted from ages of intense heat and pressure that their fossils are beyond recognition. The chances of discovering new, undistorted Archaean sediments lessens as Earth's surface geology is explored more completely and, with the loss of these sediments, our odds of finding older fossils diminishes too.

Mechanisms of stromatolite growth are also noteworthy. Their mounds—which range from narrow columns, mere centimeters across, to vast domes many meters wide and just as high—are created as layers of mud are trapped in calcium carbonate (a substance otherwise known as limestone, the skeletal material of choice for many organisms). Stromatolites produce calcium carbonate as a by-product of converting sunlight to energy and, although their mud and limestone mounds are crude structures compared to the shells and skeletons in other life-forms, they were the first examples of living organisms using rigid body frameworks to enhance their survival chances. Stacking sediment and limestone into towers lifted Archaean bacterial cells from the sea bed, bringing them closer to sunlight, reducing their chances of burial by current-swept debris, preventing neighboring mounds from overgrowing them, and enhancing their resistance to turbulent, stormy seas. Collections of stromatolites formed the first reefs, structures that are essential foundations of certain coastal habitats even today.

Cutting a stromatolite open reveals a banded structure representing periods of fast and slow growth, reflecting cycles in seasons and light availability. Variation in column thickness provides a record of the drama of a growing colony struggling against sediment being washed over the cells. In calm periods they thrived, the colony growing strong and spreading wide to create robust, thick columns. But large influxes of sediment could bring disaster, burying most of the colony and forcing the cells closest to the surface to begin the building process again, if they survived. A stromatolite fossil not only provides a record of ancient life, but also tells us much about ancient environments and climates.

The Ediacaran Biota

MULTICELLULAR LIFE HAS ARISEN INDEPENDENTLY SEVERAL times. Plants, animals, and fungi developed the capacity to form bodies and discrete tissue types independently of one another, and in all likelihood other forms of now-extinct multicellular life existed in Deep Time, particularly in the Proterozoic. The fossilization potential of these early multicellular organisms is extremely low however, and in any case most rocks from the Proterozoic Eon are unlikely to preserve fossils on account of their deformation. Despite this, a few tantalizing fossils of early, macroscopic multicellular life, such as the 2.1-billion-year-old Gabonese Francevillian biota, have snuck through to the modern day. These recently discovered marine organisms—perhaps best interpreted as amoeba-like colonies arranged into sheetlike lobes, chains, and tubes—are known from only a single locality, but they give hope to those of us wanting to know more about the early evolution of life. It's reassuring to know that some surprises still await us in our increasingly well-surveyed geological record.

A more detailed and well-studied suite of macroscopic, complex organisms is the Ediacaran biota. These organisms, which occur as fossils in rocks of Australia, northern Europe, Canada, Russia, and southern Africa, thrived between 600 million and 542 million years ago. Some survived into the Cambrian Period before becoming extinct 510 million years ago. Distinctive assemblages of Ediacaran species are recognized, seemingly correlating to different habitats such as coastal and deltaic settings, river systems, and deep marine environments.

What sort of organisms the Ediacarans were, and how they lived, are matters of ongoing discussion. Their anatomy is more complex than anything known on Earth until this time, but, alas, most Ediacarans were entirely soft-bodied and fossilized only as sediment impressions, limiting our understanding of their bodies and tissues. Most species have a pleated or ribbed appearance which is often described as "quilted." Clear morphological distinctions between different species suggest they were well differentiated ecologically. Frondlike species, such as *Charnia*, which could grow up to 2 m in length, seem to have been particularly diverse and abundant. *Dickinsonia* is generally thought to be a low-lying mobile mat, while *Kimberella* resembles a sluglike creature, its association with trace evidence of movement and apparent "head" anatomy adding credence to this interpretation. Some species were clearly burrowers, having left evidence of tunneling traces in ancient sediments. Microbes and different forms of algae were also abundant in Ediacaran times, and together with the Ediacaran organisms they might have formed some of the first complex ecosystems, the precise workings of which are still largely unclear.

How Ediacarans are related to living species also remains an open question. Might they have affinities with some of the earliest forms of animals, such as corals? Are they macroscopic, highly evolved planktonic organisms? Bizarre bodies of fungi or algae? Sophisticated bacterial mats? Or none of the above—might they be an entirely independent, now-extinct branch of multicellular life without any close living relatives? The growing consensus seems to be that Ediacarans are not a single evolutionary branch of their own, but a collection of species of varying ancestry. Some may be unique experiments in multicellular life, but others might be related to animal types still in existence today, including mollusks (the group that includes slugs, clams, and squids), cnidarians (the coral/jellyfish clade), and sponges, as well as some grades of early animal evolution that defy easy categorization. Research into Ediacarans is still in its relative infancy, the significance of their fossils being appreciated only in the 1950s and many key discoveries being made only in recent decades. Much work lies ahead to better understand the evolutionary relationships and lifestyles of these fascinating organisms.

The Cambrian Biota

THE FOSSIL RECORD BEGINS IN EARNEST IN ROCKS FROM THE Cambrian Period, the first interval of a 541-million-year division of time known as the Phanerozoic Eon (meaning "visible life"—a reference to the ubiquity of fossils in rocks of this time). This sudden change in fossil abundance reflects the widespread evolution of mineralized animal tissues: shells, spines, skeletons, and mouthparts composed of hard-wearing biominerals like calcium carbonate and hydroxyapatite. These hard parts are more readily fossilized than soft tissues like muscle, internal organs, or skin because they rot slowly, are less tempting to scavengers, and are more resistant to physical wear and tear. Historically, the sudden abundance of fossils in Cambrian rocks was thought to reflect an "explosion" of animal diversity—a rapid genesis and evolution of complex organisms after eons of nothing but microbes—but a more nuanced interpretation has replaced this today. Genetic data, geochemistry, and rare fossils indicate that many animal lineages were established several tens or even hundreds of millions of years before the Cambrian, but their lack of mineralized body tissues precluded them from entering the fossil record on all but the rarest occasions. Animal life certainly diversified in the Cambrian, but it was building on a foundation of lineages established before the Phanerozoic. This was still an explosion of diversity in many respects, but perhaps not the "ground zero" of animal evolution we've historically assumed it was.

Many Cambrian animals defy easy interpretation, but careful research has revealed that many are related to animal lineages we know today. Arthropods—the group that includes insects, crustaceans, and arachnids—were abundant and diverse at this time, scuttling and swimming around Cambrian habitats in a great variety of forms. The famous trilobites were among them, but comprised only a fraction of Cambrian arthropod diversity—a status they would later overturn. Mollusks were also present and had already differentiated into many of their major lineages. Reefs were being constructed by sponges and mineralizing microbes. Corals were present in the Cambrian but were not yet major components of reef structures—they would not adopt this role until the next geological period, the Ordovician. Small eellike creatures existed with basic, rodlike cartilaginous skeletons. These creatures would eventually give rise to vertebrates, animals with spinal cords and internal skeletons.

A major factor in the development of mineralized body parts was the need for firm tissues for defensive and offensive use: body armor to deter predators, jaw parts to process prey, and claws and hooks to seize food or fight with. Shelled animals from the close of the Ediacaran Period show that an evolutionary arms race between predators and prey was already underway before the Cambrian, and the Phanerozoic saw this develop further as animals evolved armor en masse to resist ever-tougher and more powerful mouthparts, pincers, and rasping equipment. Among the most formidable predators of this interval were the anomalocarids, free-swimming relatives of arthropods that grew up to a meter long. These unusual animals were equipped with a pair of gripping arms, two insect-like compound eyes, and a ring of slicing mouthparts. Two rows of segmented fins along their bodies enabled them to float around reefs in the manner of cuttlefish, seizing prey with their arms before crushing and slicing it with their jaws. Exactly what anomalocarids ate is uncertain, though various arthropods and soft-bodied organisms were probably common prey. Some species, however, had elongated arms equipped with fine combs, which they may have used to snag small prey and particles of food in the water column, rather than hunting down larger prey on the seabed.

Trilobites (Ordovician)

TRUNDLING ACROSS THE SEAFLOOR FOR 270 MILLION YEARS WERE the trilobites, creatures familiar to anyone who searches for fossils among Paleozoic rocks. Trilobites lived in seas all over the planet and left an extensive fossil record of over ten thousand species, most of which stem from their heyday in the first half of the Paleozoic. Their later history was less successful: trilobite diversity dwindled in the Devonian to leave just one group, the proetids, hanging on until their final demise at the end of the Permian.

There is little doubt that trilobites are arthropods, but exactly how they are related to other members of this group is not clear. It might seem surprising that the origin of trilobites remains mysterious given their excellent fossil representation and long research history (scholarly interest in trilobites dates to 1698), but this is not uncommon when working on long-extinct animals. Evolutionary processes are complex, our fossil record is very incomplete, and even well-known extinct species present only limited anatomical information for piecing together their evolutionary histories. New discoveries of exceptionally preserved trilobites, as well as other early arthropods, promise to shed light on their affinities in future.

Well-preserved trilobite fossils have provided detailed insights into their anatomy, even their internal organs. Their calcified exoskeletons were divided into three units: the head (or cephalon), a series of body segments (thorax), and a shield-like terminal plate formed of fused body segments (pygidium). The head is often characterized by a bulbous central structure that looks like it should house a brain, but it actually held a stomach. The gut extended from this through the thorax and pygidium, and waste was expelled at the rear of the body. Their eyes were perched at the side of their heads and are the most sophisticated sense organs we know of among early animals. Like all arthropods, trilobites had eyes comprised of numerous tessellating lenses but, unlike insect or crustacean eyes, their lenses were not made from protein. Instead, each lens was a finely shaped calcite crystal: trilobite eyes were made of stone. This might seem primitive, but experiments show that trilobites had excellent visual acuity, seeing in sharp detail and with considerable depth of field. Variation in eye shape reveals the importance of vision to different trilobite species. Fast-moving swimmers had enormous eyes with wide visual ranges, some shallow-water species had overhanging "sun shades" to maintain their vision in strong sunlight, and low-light or burrowing species lost their eyes altogether. Further anatomical sophistication was found in trilobite legs. Each limb had two branches: a lower branch for walking or swimming, and an upper branch supporting a gill for respiration. Legs close to their mouths sometimes bore bladelike structures to assist with processing food, which was likely soft prey in many species, such as worms. When threatened, trilobites were able to roll into tight balls to protect their delicate undersides, and they are often found like this as fossils (it's best not to think about the possibility of them being buried alive to preserve in this way).

Trilobites adapted their anatomy into numerous different body shapes, sizes, and lifestyles. The most typical trilobite morph was a probably a seabed-roaming grazer or predator-scavenger, much like the large (70-cm-long), widespread species shown opposite, *Ogyginus forteyi*. But with only a few tweaks to their anatomy, very different lifestyles were possible. Species with small bodies but enormous, wide heads filter-fed with their legs while performing headstands. Some tiny-bodied, large-eyed species swam upside down through water. So-called effaced species—those that smoothed over their facial and body contours—were able to burrow. Taxa with numerous body segments and expanded, wide thoraxes may have relied on microbes living in their gills for sustenance. We are still learning how to interpret the different body shapes and adaptations of trilobites but, whatever they got up to, it clearly served them well for much of the Paleozoic.

29

The Ordovician–Silurian Extinction

THE STORY OF LIFE ON EARTH IS NOT ONE OF EVER-INCREASING biological diversity: species also become extinct. The fossil record shows that extinction is a routine part of the evolutionary process, the natural result of organisms being unable to adapt to changing conditions. The normal rate of this process is termed "background extinction," and it is generally offset by a comparable rate of speciation—the emergence of new species. But some periods of Earth's history show dramatic increases in extinction rates, where the fossil records of numerous species, and maybe even whole evolutionary lineages, terminate at a common level in geological sequences. These are indications of relatively sudden and widespread periods of environmental stresses killing off lifeforms that could not adapt quickly enough to new conditions. Many such extinction events are known from Deep Time and several—colloquially known as "The Big Five"—were global-scale mass extinctions that radically altered the course of evolutionary history. The most famous of The Big Five are the Permian–Triassic and Cretaceous–Paleogene events; with the others occurring at the Devonian–Carboniferous boundary, the Triassic–Jurassic boundary, and at the end of the Ordovician Period.

Taking place 444 million years ago, the Ordovician extinction event is the oldest global mass extinction we know of. It seems to have eliminated 85 percent of species alive at that time. Many famous Paleozoic fossil groups—the trilobites, the planktonic graptolites, the hagfish-like conodonts, and the bivalved shellfish known as brachiopods—were badly affected, though few major groups were wiped out. The Ordovician extinction was a considerable squeezing of biodiversity rather than, as with other extinctions, a curtain call for major groups of plant and animal life.

The Ordovician extinction seems to have taken place in two pulses, the first being a period of global cooling. Ordovician lifeforms were adapted to very warm, greenhouse conditions but, as the period came to a close, they found themselves in an icehouse world cold enough to allow glaciers to form over the poles. The supercontinent Gondwana—a fused landmass composed of what has since become the southern continents—was situated over the south pole at this time, allowing an ice sheet more than 6,000 km wide to grow in the southern hemisphere. The expanding glaciers sequestered ever-greater stores of the planet's water, ultimately lowering sea level by a staggering 50–100 m and emptying shallow seas around the planet. Shallow marine settings are extremely species-rich, so their loss brought heavy casualties to Ordovician biodiversity. This lowering of sea level also changed oceanic circulation and chemistry, decreasing oceanic oxygen levels while elevating concentrations of hydrogen sulfide—a substance dangerous to animals when in high concentrations. Eventually, even deep-water species that were not affected by falling sea levels risked annihilation.

But just as life began to adapt to these hostile conditions, the planet warmed, the glaciers retreated, sea level rose, and oceanic chemistry and circulation reverted to pre-extinction conditions. Though seemingly a return to normal, this sudden reversal brought on a second pulse of extinction, affecting those hardy species that had been prospering in the half-million years associated with the glaciation. It took until the middle of the Silurian Period—about 430 million years ago—for life to attain its prior diversity. But unlike other large extinction events, the postextinction biosphere returned to a similar, pre-extinction configuration instead of transforming into something radically different. Ecologically speaking, the evolution of life resumed its prior course. As we'll see later, other mass extinctions heralded more dramatic changes in the development of life on Earth.

Jawless and Jawed Fish, and "Sea Scorpions" (Silurian)

VERTEBRATES—ANIMALS WITH A STIFFENED ROD OF BONE OR cartilage supporting their bodies, surrounding a chord of nervous tissue—first appeared in the Cambrian Period. This group, to which we belong, maintained a low profile for much of the early Paleozoic Era, living in seas dominated by invertebrates. The earliest forms were probably something like modern lampreys, eellike creatures that lack a true skeleton and feed through rasping, jawless mouths. By the start of the Silurian Period, vertebrates had evolved into a number of jawless forms with distinctly fishlike body shapes, some of which bore small scales, others sporting mineralized armor. These early fish were the agnathans.

Armored agnathans include the heterostracans (middle left in the opposite illustration) and osteostracans (bottom right). They are characterized by broad shields over their faces and large, overlapping scales along their trunks and tails. Thelodonts (bottom left and middle right) were not armored, but instead covered in minute, mineralized scales. Most of these early fishes were small—perhaps 10–20 cm long—but some were giants, attaining body lengths of a meter. Even so, they were far from the most formidable animals in early Paleozoic seas and likely used well-developed senses—as indicated by fossilized details of their brain and sensory anatomy—to avoid danger, along with tough dermal tissues and spines to dissuade predators. The diversity of agnathan body shapes indicates that they were adapted for a number of lifestyles, and they represent the bulk of fish diversity throughout the Silurian, declining only in the Devonian thanks to the rise of jawed fishes—the gnathostomes.

Acanthodians were among the earliest gnathostomes, first appearing in the Silurian and reaching their evolutionary acme in the Devonian. They are more typically fishlike in anatomical detail than the agnathans, including the presence of well-developed jaws and teeth. Jawed mouths had a complex evolution, with both internal and external anatomy being co-opted for biting actions. Jawbones evolved from structures that ancestrally supported a gill, while teeth share properties with structures that once covered skin. Being able to bite, rather than merely rasp or suck, was a major development for vertebrates that was critical to their later occupation of predatory roles in Devonian seas. Invertebrate predators, such as crustaceans and scorpions, rely on different structures to apprehend and process food, while vertebrate mouths permit the dismembering, immobilization, and devouring of prey with one organ. This enables a more efficient and streamlined body plan, as well as the capability to develop and optimize one organ, rather than several, for predation and ingestion. The continued success of animals like sharks—which first appeared in the Silurian, if not the Ordovician—shows the potential inherent in mounting a powerful jaw on the front of a streamlined, swimming vertebrate body.

Swimming above these early fish is a eurypterid, a giant arthropod that belongs to the same group as spiders, horseshoe crabs, and scorpions. These "sea scorpions" were major predators in early Paleozoic seas, and they have an evolutionary range from the Ordovician to the end of the Permian. They were particularly abundant and diverse in the Silurian and were able to tear into prey with large pincers. Though many eurypterids were well adapted to walking and likely patrolled the sea floor (as well as freshwater settings, and perhaps even terrestrial environments in their later evolution), some had reduced limbs save for a pair of paddle-like swimming appendages that propelled them through water. At up to 2.5 m long, the biggest eurypterids were among the largest predators of the early Paleozoic as well as the largest arthropods of all time. They must have been intimidating animals to witness in life. But the heyday of eurypterids was fleeting. Their diversity declined early in the Devonian and they lost their status as arch predators, never to regain it. The cause of this is unknown, but competition from jawed fish may have been a factor.

Plants Colonize the Land (Silurian)

FOR BILLIONS OF YEARS THE CONTINENTS WERE BARREN, LIFE-less masses drifting over Earth's mantle. These stark landscapes offered few opportunities to early life, having little in the way of food or nutrients for animals and only limited options for habitation by microbial organisms. For life to escape the seas and colonize the land, continental habitats needed to offer environments with abundant amounts of chemical energy and sufficient nutrients to build and maintain organic matter. The organisms that rose to this task, and changed the evolution of life on Earth forever, were the vascular plants.

Both animals and plants were experimenting with terrestrialization throughout the first half of the Paleozoic. Fossil footprints show small arthropods making quick trips out of the water as early as the Cambrian, genetic studies suggest that bryophytes (mosslike plants) existed on land in the same period, and fossils of the earliest land plants date to the Ordovician. But it was not until the late Ordovician and early Silurian that land plants were anything more than small mossy carpets, a change marked by newly evolved plants raising parts of their bodies from the ground via short stalks. They could do this thanks to an important innovation: vascularity. Vascular plants are distinguished from bryophytes by the presence of internal tubes that ferry water and nutrients around their bodies. The products of photosynthesis (the process of converting light to chemical energy) are carried in tubes known as phloem, while xylem, tubes made of a hard substance called lignin, transport water. Lignin is also the material that, in sufficient quantity, makes plant tissues woody and rigid.

The vascular plants colonizing mid-Silurian landscapes were small, no more than a few centimeters tall, and had a relatively simple vascular system. Their most famous genus, *Cooksonia* (shown opposite, bottom left), was a globally distributed branching plant that lacked leaves but bore swollen sporangia at the tips of its stems for distributing reproductive spores. *Cooksonia* and other early land plants lacked roots, these having no purpose on an Earth without deep soils. As a mix of organic and rock matter, soils could not form in earnest until large quantities of biomass had accumulated on land, and the soils encountered by Silurian plants were very thin, likely just millimeters or centimeters deep. It was not until the Devonian that soils of modern depths and nutrient quality were achieved.

The utility of vascularity was quickly realized by land plants. An internal transportation system allowed their bodies to grow large and differentiate their tissues, exchanging nutrients between organs specific to photosynthesis and organs adapted for absorbing water. The lignin in their tissues allowed for the development of more complex and intricate anatomies, including long branches, stabilizing grapples, and anchoring roots. By the end of the Silurian Period another plant genus, *Baragwanathia*, demonstrated many of these features, and compared to *Cooksonia*, it was a giant of many tens of centimeters in height. Plant communities from the early and mid-Devonian included multiple species further capitalizing on these adaptations, forming small, shrubby forests. The first trees and seed-bearing plants appeared halfway through the Devonian, about 385 million years ago—just 50 million years after *Cooksonia* began growing tall.

The development of land plant communities had many significant impacts on Earth's atmosphere, geology, and biosphere. Plants cooled the planet, absorbing atmospheric carbon dioxide (a potent greenhouse gas) as they photosynthesized and locking it into biological systems, as well as slowed erosion through roots binding rocks and soils. This not only changed how landscapes were shaped, but also slowed nutrient cycling between land and sea. New, plant-dense habitats also created new environments for land animals to live in. The first to exploit these were the arthropods, creatures that followed plants onto land in the Silurian. Arthropods thrived in this world of early land plants, untroubled by other types of land animal for millions of years.

Titanichthys *(Devonian)*

MANY OF THE LARGEST CREATURES IN OUR OCEANS TODAY FEED on some of the smallest: plankton. They obtain these tiny prey items using filtration mechanisms that preclude the need for precision prey capture: they simply aim their open mouths at volumes of plankton-rich water and, after engulfing water and food alike, sieve their food using filters or combs. One mechanism for this is ram feeding, swimming with mouth agape to strain plankton through fine, comblike gill rakers. Manta rays and the largest sharks (the basking and whale sharks) use this method. Rorqual whales employ a strategy known as lunge feeding, where mouthfuls of food are sieved from water forced through sheets of brushlike baleen lining their jaws by contraction of a huge, powerful throat. The rorquals include the largest animals to have ever lived, including the fin and blue whales (up to 33 m long).

Growing large on a plankton diet is not a modern biological innovation. During the Jurassic and Cretaceous Periods the principle large ram feeders were the pachycormids, a globally distributed lineage of bony fish that matched the biggest living sharks in size. The largest was the Jurassic species *Leedsichthys problematicus*, which averaged between 7 m and 12 m in length but on occasion might have reached 15 m. The closely related *Bonnerichthys gladius* was also a large animal, perhaps achieving 5 m in length. Some members of the sand shark lineage (odontaspidids) may have also experimented with planktivory toward the end of the Cretaceous. Curiously, for all their longevity, frequent development of giant size, and anatomical diversity, the Mesozoic marine reptiles do not seem to have capitalized on this niche in any great capacity.

But the pioneer giant planktivore was the Devonian placoderm *Titanichthys agassizi*, pictured here. *Titanichthys* was among the last of a lineage known as the arthrodires, a diverse and abundant group of armored fish that occupied numerous roles in marine ecosystems throughout their fifty-million-year history. They were, for the time, the biggest animals on Earth, with both *Titanichthys* and the predatory arthrodire *Dunkleosteus terrelli* reaching around 6 m in length. Complete remains of these giants remain elusive because their skeletons were mostly composed of cartilage, which rarely fossilizes, leaving only their relatively robust skulls and armor to persist in the rock record. Fossils of *Titanichthys* have been found in rocks across the United States, Europe, and Africa since the late 1800s, but much remains uncertain about its paleobiology. This includes aspects as fundamental as the number of species: seven *Titanichthys* species have been named (five of which are from the area that is now the United States), but only three are represented by relatively good fossils. We do not yet appreciate how much *Titanichthys* individuals differed from one another or how their proportions changed with growth, and the provenance of many *Titanichthys* fossils—exactly where they were found, and in what rock layers—is poorly recorded. All these factors complicate assessments of their diversity.

It's also only in recent years that we've developed a reasonable understanding of *Titanichthys* skull structure and found stronger evidence of a ram-feeding lifestyle. Unlike *Dunkleosteus*, which had famously robust jawbones shaped into bladelike biting surfaces, *Titanichthys* had smaller, relatively slender jawbones lacking blades or other toothlike structures. It also had small eyes and a peculiarly downturned lower jaw that likely expanded the circumference of the open mouth. These are interpreted as adaptations for ram feeding, and we should probably picture *Titanichthys* obtaining its food by swimming through plankton-rich seas with its mouth open, establishing a lifestyle that numerous ocean giants would emulate in the millions of years to come.

Lakes of the Early Carboniferous

THE FRESHWATER FISH OF THIS EARLY CARBONIFEROUS SCENE represent the continued success of fishes through the late Paleozoic Era. Though some early fish lineages, such as the placoderms, were now extinct, the sharks, ray-finned fishes (actinopterygians), and lobe-finned fishes (sarcopterygians) went from strength to strength. One group of sarcopterygians—the rhipidistians—had adapted to live in estuaries, rivers and lakes by the Devonian. This transition would be important for the evolution of limbed fish (including ourselves) throughout the Devonian and Carboniferous Periods. Once adapted to life in freshwater, the rhipidistians split into two major groups: the lungfishes (Dipnoi) and the tetrapodomorphs—the fishes that gave rise to limbed vertebrates that could walk on land.

Carboniferous lakes and rivers were occupied by some of the largest freshwater fish of all time: the rhizodonts (center in opposite image). These tetrapodomorphs first appeared in the Devonian, and they became extinct in the late Carboniferous. Some were moderately sized—less than a meter long—but several species reached 5–7 m: the same length as a modern great white shark. They must have been formidable animals, having stress-resistant jaws with two rows of teeth, 20-cm-long fangs, and bodies covered in tough, platelike scales. The tissues of their teeth were arranged in convoluted rings that increased their strength, allowing their use in forceful stabbing motions even though they were only weakly anchored to the underlying jawbones. These features cast rhizodonts as powerful predators that surely ate other large vertebrates. Studies of rhizodont teeth imply that strong armor, rather than large size, was the best defense against them.

Rhizodont forefins were broad, stiff, and capable of a wide range of movement. They seem adapted to facilitating rapid changes in direction. But rhizodont bodies were otherwise long tubes, with their dorsal, anal, and pelvic fins situated far down the body, effectively being incorporated into the tail fin. This arrangement seems better suited to powerful, rapid acceleration than to high speed, and their unusually flexible vertebral columns allowed them to flex not only sideways—as we typically associate with fish—but also up and down. This body plan must have made rhizodonts agile, maneuverable swimmers despite their size, and perhaps also assisted with dismembering large prey: it's hypothesized that rapid, forelimb-aided shaking or tail-aided corkscrewing may have been used to rend big animals into consumable pieces. The enhanced sensory system that ran across rhizodont faces and scales probably allowed these fish to detect prey and obstacles even in murky water. In short, if you ever have the chance to visit the Carboniferous, you needn't take your swimming gear.

Rhizodonts lived alongside lungfishes (middle right) and *Gyracanthus* (left and right). Lungfishes have an extensive fossil record that peaks in the Triassic, the apex of their diversity and distribution. They are most famous—and from our perspective, most evolutionary-significant—for developing air-breathing lungs that are the precursors to our own. They are not the only fish capable of absorbing oxygen from air, but their lungs allow them to exist outside of water for sustained periods, a habit that some modern lungfish (such as the spotted lungfish, *Protopterus dolloi*) use to weather droughts in mucus-lined shallow burrows. *Gyracanthus* is a mysterious acanthodian known almost entirely from ornamented forelimb fin spines and shoulder bones. Many fossils of this large (perhaps 1.25 m long) fish are known from Devonian and Carboniferous rocks, and these give some indication of a small head and triangular body section. Further details of its full form remain elusive, however.

Pulmonoscorpius, *Giant Scorpion of Ancient Scotland (Carboniferous)*

READERS WITH A PHOBIA OF BUGS, SPIDERS, AND OTHER SPECIES with more than a sensible number of legs would not enjoy hikes in Carboniferous landscapes. Though the swamplands and the lush flora of giant scale trees, early conifers, horsetails, and ferns would make for excellent walking country, the presence of extremely large arthropods would activate bug-based fears in all but the steeliest individuals. Large, semiaquatic eurypterids scurried between ponds while 2-m-long millipede relatives, *Arthropleura*, scuttled through swampland in search of nutritious plant matter. Large dragonfly-like insects known as meganisopterans (of which the 70-cm-wingspan *Meganeura* is the most famous) buzzed overhead, and *Pulmonoscorpius kirktonensis*—a 75-cm-long Scottish scorpion (shown opposite)—stalked smaller arthropods and our own tetrapod ancestors. Scorpions have a long evolutionary history beginning in the Silurian, and *Pulmonoscorpius* is the largest to have ever lived.

Though the most famous of the giant arthropods are Carboniferous in age, their experiments with gigantism were a long-lived phenomenon that extended into the Permian. Species of mayfly, the extinct paleodictyopterans, and several types of wingless insects also attained elevated body sizes in these periods. Exactly what facilitated late Paleozoic arthropod gigantism remains a topic of contention. The traditional explanation is that higher levels of atmospheric oxygen ("hyperoxia") enhanced animal respiration, permitting more powerful muscle activity and the capacity to support a larger, heavier skeleton. An alternative spin on the hyperoxia idea is that smaller arthropods found elevated oxygen levels somewhat toxic, promoting the evolution of giants that were better able to cope with high oxygen levels. Both ideas have some support from experiments on living insects.

However, there are several counterpoints to the hyperoxia hypothesis that are not easily shaken off. Among the most critical is that most Carboniferous and Permian arthropod species were not gigantic, and it has not been demonstrated that arthropods of this time were, on average, generally larger than at other times in history. Thus, we've yet to determine if we're being misled by a few exceptionally large species (which would likely reflect specific adaptive or ecological innovations of a few lineages) or a global trend of arthropod gigantism (which would be more consistent with environmental conditions elevating arthropod size limits). It is also clear that hyperoxia is not essential for large terrestrial arthropods, thanks to the modern, strongly terrestrial coconut crab (*Birgus latro*). This 4-kg giant, which has a leg span of 90 cm, remains on land via a sophisticated lunglike organ adapted for breathing air, a structure somewhat analogous to the "book lungs" of spiders and scorpions. We know that *Pulmonoscorpius* had a similarly adapted respiratory mechanism and, if it functioned as well as its coconut crab equivalent, *Pulmonoscorpius* may have been capable of absorbing all the oxygen it needed even at modern-grade oxygen levels. While this does not explain gigantism in late Paleozoic insects (they rely on a less-efficient gas exchange mechanism formed from tubes invading their bodies), it demonstrates that not all arthropods need unusual environmental factors to evolve to huge size. The lack of vertebrates on land until the late Paleozoic is another factor we must consider: these animals were surely predators of large arthropods, as well as competitors for their food, so their absence in the early Carboniferous may have created relaxed ecological conditions for arthropods to experiment with large body size. More work is needed to explain why some Carboniferous and Permian arthropods became so gigantic, but we should note that the ideas discussed here are not mutually exclusive. Evolution and adaptation are complex, and it's entirely plausible that hyperoxia, anatomical innovation, and relaxed ecological pressures in Carboniferous landscapes played complementary roles in boosting the body sizes of certain ancient arthropods.

41

The Tetrapods Invade the Land (Carboniferous)

LIFE'S INVASION OF LAND TOOK OVER ONE HUNDRED MILLION years to complete. We have already met some of the pioneering land plants and arthropods, but we've yet to cover the final stage in this process: when fish hauled, crawled, and slid from the water onto land, eventually evolving to walk and breathe in terrestrial settings. The evolution of land vertebrates, or tetrapods, is widely regarded as one of the most significant steps in vertebrate evolution, and it was a far-reaching event for the Paleozoic biosphere.

The broad strokes of tetrapod evolution have been known for some time, but only in recent decades have we begun to uncover the details and specifics of this major evolutionary event. The creatures shown in this early Carboniferous scene represent some of the first tetrapod-like animals, species that had limb-like fins but were not yet capable of carrying themselves over land in a true walking gait. They likely propelled themselves along on their bellies, using their appendages to push and pull their bodies across lake margins and riverbanks. The landscapes they entered were not the richly forested swamps of the latter Carboniferous Period, but the less densely vegetated precursors to these environments.

Different grades of tetrapod evolution are shown in this illustration. Animals such as the unnamed species on the far left of the scene and the fishlike whatcheeriid on the right represent the epitome of vertebrate land evolution at this time. They probably still spent much of their time in water, feeding on fish or aquatic invertebrates, but their deep bodies and relatively powerful limbs enabled them to enter terrestrial settings when they desired. Seeking safety, exploiting new foraging opportunities, and travelling between aquatic habitats could have been catalysts driving this behavior. It's likely that their forays onto land did not require novel gaits, because many near-tetrapod fish—both living and extinct—can crawl or walk with their fins even when they're submerged in water. Thus, the major challenge for early tetrapods was not the kinematics of locomotion itself, but carrying their bodies away from the supportive medium of water. Fossils show that tetrapods quickly developed longer and stronger legs once they were experimenting with land-based locomotion, allowing for longer bouts of terrestrial behavior and more efficient means of supporting themselves away from water.

The large, eellike species in the center of the scene is a colosteid, a long-bodied animal with diminutive limbs and a flattened body profile. Although they're somewhat more closely related to the true tetrapods than to the other animals shown here, colosteids were likely very sluggish land creatures and may have reverted to an entirely aquatic way of life from a semiterrestrial ancestor. The fact that colosteids split from the main tetrapod line and reversed the general trend of their evolution is a great example of the complexity of natural selection. When discussing grand evolutionary events such as vertebrate terrestrialization, we can give the false impression of species moving toward goals, striving for a generationally distant biological optimum. But lineages like the colosteids show that evolution is far more opportunistic: organisms make the best of whatever works for them at the time, not what might be ideal in one hundred generations. It seems that, for colosteids, abandoning experiments with terrestriality in favor of returning to a more aquatic way of life was more successful than following the adaptive trends of their relatives. As we learn more about the early history of Tetrapoda, we see that it is a story not just of fish leaving the water, but of vertebrate groups adapting to life at the water's edge. Only some of these groups would seek further evolutionary novelty by pushing further inland.

Platyhystrix, *a Sail-Backed "Amphibian" (Carboniferous–Permian)*

BY THE END OF THE CARBONIFEROUS, TRUE TETRAPODS WERE roaming the landscapes and swampy waterways of the world. One group, the amniotes, had already split from the rest of the tetrapod lineage and would, aside from the need to drink, sever all essential ties with water. Their final hurdle to an entirely land-based existence concerned reproduction: while the first tetrapods were competent land animals, they still had to lay eggs in aquatic settings. This was overcome with the evolution of the amniote egg: a tough-shelled, desiccation-resistant capsule containing enough nutrients to support and nourish a developing embryo, as well as an efficient gas exchange system to supply oxygen and expel carbon dioxide. Now fully able to exploit terrestrial settings, the amniotes gave rise to the two great lineages of land vertebrates: the diapsids (reptile-line animals) and synapsids (mammal-line animals).

Another major tetrapod lineage arose in the Carboniferous: the temnospondyls. These diverse animals are what we think of as "prehistoric amphibians," although their relationship to living amphibians is debated. Temnospondyls were mostly somewhat salamander-like species that lived in or around water, but some species were more terrestrially adapted. They attained a range of body shapes and sizes that permitted occupation of a number of ecological niches. Many were crocodile-like, snatching fish or ambushing shorebound prey with broad, flattened heads filled with teeth. Some spent much of their time on land and may have competed with carnivorous amniotes for prey, while others became giant marine predators. Like living amphibians, they had larval stages where they bore long gills emerging from the back of their heads, and we sometimes find these in their fossils. Temnospondyls were a long-lived lineage that was particularly speciose in the Carboniferous, Permian, and Triassic Periods and, although their diversity and abundance

lessened throughout the Mesozoic Era, they persisted in some parts of the world until the Early Cretaceous (and maybe longer, if living amphibians are their descendants).

Among the most remarkable of the temnospondyls was *Platyhystrix rugosus*, a large (about 1 m long) species from latest Carboniferous and earliest Permian rocks of the southern United States. The sail of this animal is most striking and was formed of elongate extensions of its vertebrae fused to bones in its skin. *Platyhystrix* is a dissorophid, a temnospondyl lineage characterized by strong terrestrial capabilities and extensive body armor. They bore long, robust limbs; well-developed limb girdles; a strong spinal column reinforced by bones in the skin; and a high level of skeletal ossification: attributes which equipped *Platyhystrix* for sustained periods of walking and running. Dissorophid skulls were powerfully built and bore numerous conical teeth both around the jaw margins and within the mouth. These leave little doubt of their predatory habits, and their well-developed eye and ear anatomy likely outfitted them with suitable senses to locate prey as well as detect danger. With these adaptations and their widespread distribution, dissorophids may have been important land predators during the late Carboniferous and the early Permian. Their carnivorous diet is confirmed by bite marks and shed teeth associated with fossils of Permian synapsids: this temnospondyl not only competed with amniotes for food, but actually made meals of the competition.

Several dissorophids have sails, but they are not ubiquitous among the group. The function of the sails is not clear, but several ideas have been suggested: to regulate body temperature, to reinforce the spinal column, as defense structures, or as sociosexual display devices. We'll return to the issue of sailed animals soon.

Harlequin Mantella Frog and Other Amphibians (Holocene)

MODERN AMPHIBIANS—THE CLADE LISSAMPHIBIA—CAN BE DIvided into three major lineages: frogs, salamanders, and caecilians. Frogs are overwhelmingly the most speciose group, representing something like 90 percent of all living lissamphibians. In contrast to the large, robust tetrapods we met on previous pages, lissamphibians are generally small-bodied animals with delicate skeletons. This means they do not fossilize easily and have an accordingly patchy fossil record. Long-standing disagreements over their relationships to other tetrapods complicate our understanding of their early history. Some models suggest Lissamphibia are a subgroup of the temnospondyls (in which case temnospondyls did not go extinct in the Cretaceous), while others posit that the lepospondyls, another group of early amphibious tetrapods, are their true evolutionary home. A third model splits these ideas, with some modern lissamphibians arising from temnospondyls, and others from lepospondyls. A wholly temnospondyl origin is currently preferred, but it does not answer all the questions we have over lissamphibian evolution, such as which fossil temnospondyls are their closest relatives, or whether they evolved in the Permian or the Carboniferous.

Lissamphibians are an overlooked tetrapod group in modern culture. Being mostly small creatures that inhabit dark, damp environments, and that struggle to live in close association with humans, especially in polluted ecosystems, they fly under our collective radar far more than birds, reptiles, or mammals. Our general unfamiliarity with amphibians is our loss. These remarkable animals offer some truly spectacular takes on tetrapod anatomy: peculiar skeletons, a suite of remarkable adaptations related to breathing (including being able to breathe through their skin; using a throat-pump to initiate respiration; and, in some species, having no lungs at all), life cycles that involve transitioning from gilled swimming larvae to air-breathing adults, and the ability to completely regenerate lost body parts. They are also amazingly diverse in anatomy and lifestyle. Frog hind-limb and pelvic anatomy is strongly specialized to leaping, but some species also excel at climbing, digging, and swimming. Salamanders are not only newt-like in form; they also exist in eel-like varieties that are strongly specialized for aquatic lifestyles. Caecilians are perhaps the strangest of all, being limbless creatures that are at home underground or in stream substrates, and resemble worms or snakes depending on their size. Fossils show that lissamphibians occupied similar body forms in the deep past: dinosaurs would not be startled by modern-looking frogs, salamanders, or caecilians.

Today, amphibians face real adversity. Drastic falls in amphibian populations have been recognized in thousands of species across the world since the 1950s, and their demise continues today as severely as ever. Hundreds of species are now classed as critically endangered following decades of continued pressure on amphibian communities, and many species now only survive in captivity. The necessity for both land and water in amphibian life cycles means they are especially vulnerable to environmental modification, placing climate change (and the associated rise in ultraviolet light radiation) and habitat degradation as major drivers of population loss. Diseases (possibly spread by travelling humans and exacerbated by habitat changes) and the introduction of foreign predators into amphibian habitats also play a role. Wild collecting for the pet trade has reduced some wild amphibian populations to critical levels and has even—as with the famous gilled salamander, the axolotl—rendered them extinct in the wild. In many cases, such as with the harlequin mantella frog (*Mantella cowani*, shown here), amphibian species have become endangered before we even have time to understand their biology. The unavoidable fact of amphibian conservation is that this great branch of tetrapod evolution will be decimated within our own lifetimes unless significant and immediate efforts are made to conserve them.

Caseids: The Land Vertebrates Declare War on Plants (Permian)

COMPARED TO CARNIVORY, WHERE MEALS TEND TO BE FLIGHTY, armed, or angry (or all three), herbivory may seem like an adaptive walk down easy street. After all, what is difficult about eating plants? They're abundant, accessible, and mostly inoffensive to the body parts of the individuals devouring them. But herbivory is actually a much tougher lifestyle than it first seems. The flesh of our fellow animals is relatively easy to digest and rich in nutrients, on account of it being in the same chemical format as our own tissues. Plants, in contrast, are composed of a different cellular makeup that makes them tough to digest and nutrient-poor, to the extent that many living herbivores have to supplement their diet with animal tissues on occasion. It takes a number of significant adaptations to eat a plant-based diet: abrasion-resistant mouthparts that can reach and crop plant tissues, a digestive mechanism that can break into tough cells, a gut that can hold huge quantities of bulky vegetative matter and absorb its nutrients, and a body capable of supporting and moving a relatively bulky torso. Herbivory may look easy, but it's actually more complex and physiologically challenging than carnivory.

It took tetrapods several million years to develop lineages capable of eating plants. Among the first major successes in this niche were the caseids, a group of large-bodied Carboniferous and Permian amniotes that look like reptiles but are in fact more closely related to mammals. Caseids are synapsids, the major branch of tetrapod evolution that contains mammals and our ancestors. Our early synapsid relatives were the dominant terrestrial vertebrates in the late Paleozoic, transforming from superficially reptile- or amphibian-like creatures in the Carboniferous to a diverse range of increasingly mammal-like animals through the Permian. Caseids were some of the first synapsids to become widespread and abundant, living across North America, Europe, and Russia for much of the Permian and outlasting several other early synapsid attempts at herbivory. They were among the largest animals on land in the Permian. Some species, such as the completely known *Cotylorhynchus romeri*, regularly attained lengths exceeding 3.5 m and masses of 300 kg, but other species—such as *C. hancocki*, shown here—reached 5–6 m long and weighed half a tonne.

The first caseids were likely carnivorous, but the group quickly switched to dedicated herbivory, becoming barrel-chested, small-headed animals with enormous, sprawling limbs. Their small heads look ridiculous at first glance, but they are not so different in proportion to those of herbivorous tortoises or sauropod dinosaurs. If herbivores have a large enough gut mechanism, their mouths simply have to harvest plant matter, negating the need for a large head and robust teeth for chewing. Caseid jaws were equipped with a dentition similar to that of iguanas, along with smaller teeth across the roof of their mouths and powerful tongues (the latter not being directly preserved, but evidenced by robust bones that anchor throat and tongue tissues). Collectively, these anatomies allowed caseids to shear and rip away plant material before sending it to their expansive guts for a long digestive process. Their gut capacity and dentition seem suited to high-fiber vegetation, one of many paleobiological details indicating a fully terrestrial lifestyle for these creatures.

Caseid limbs are extremely stout, and their enormous hands and feet were equipped with huge, pointed claws. Much of their limb anatomy has been regarded as an adaptation for weight-bearing at large size, but the fact that even small caseids have robust limbs suggests a function beyond supporting huge bodies. They were likely excellent diggers, their robust forelimbs and powerful hands perhaps being used to uproot plants to access their nutritious roots, or perhaps to dig burrows. The idea of such large animals excavating burrows may seem peculiar, but we know that similarly sized ground sloths and bears are capable of such feats, and it would be remiss to preclude this behavior for these evidently mighty extinct animals.

Dimetrodon *(Permian)*

THE MOST FAMOUS AND CHARISMATIC OF ALL EARLY SYNAPSIDS is surely *Dimetrodon*, an animal best described as an alligator–bull terrier cross with a penchant for sails—the ultimate retro prehistoric accessory. *Dimetrodon* is a sphenacodontid, a group of European and North American predators that evolved in the Carboniferous and went extinct in the mid-Permian. They represent a grade of synapsid evolution that was somewhat more mammal-like than a caseid, but was still superficially reptile-like in many respects. Many sphenacodontids were small (just over half a meter long) but some, including several *Dimetrodon* species, attained total lengths of 4.5 m and body masses of 250 kg. Large sphenacodontids were probably top predators in many terrestrial Permian food webs.

Dimetrodon occurred in Permian landscapes of the United States and Germany, and up to twenty species have been recognized from its abundant fossils. Modern assessments of its diversity suggest that only thirteen or so of these proposed species are valid, however. The taxon illustrated here is the large Texan species *D. grandis*, one of the best-represented and studied taxa. Fossil bone chemistry, which reflects some details of what animals were eating and where they obtained their food, suggests that *Dimetrodon* was primarily a terrestrial predator but probably not a fussy eater. Small individuals likely ate insects and small vertebrates, while larger animals preyed on fish, amphibians, and other synapsids. These interpretations are supported by possible *Dimetrodon* stomach contents and bite marks on prey species, including the "boomerang-headed" lepospondyl *Diplocaulus* and freshwater sharks. In the adjacent scene, a *Diplocaulus* has been pulled from its burrow before being eaten by a *Dimetrodon* family.

The function of the *Dimetrodon* sail is a subject of much intrigue. Sails were common to several sphenacodontid species and several other early synapsids, but they are not so universal to imply that they were physiologically essential. Indeed, large sail-less sphenacodontids, such as *Sphenacodon,* lived in the same time and place as *Dimetrodon*, which rules out environmental conditions as promoting sail development. Sails might have imparted practical benefits, such as helping their owners warm up in direct sunlight or cool down in a breeze. But they also have drawbacks: significant resource investment is required for their growth and maintenance, they restrict mobility when the animal is moving through cluttered habitats, and they make their owners conspicuous to predators. Studies of *Dimetrodon* sail growth suggest that they grew much faster and larger than expected for a device required to regulate body temperature, instead matching the growth rates of sociosexual display structures in living species. This implies a social selection pressure—the sail as a display device to attract mates or deter rivals—over a mechanical one, a finding that also matches the somewhat random distribution of sails among Permian animals. The display structures of living animals often have the same traits—a complex, seemingly erratic occurrence across evolutionary lineages, and detachment from fundamental physiological functions—and we can assume some elaborate anatomies of past species served the same purpose.

Investigations into *Dimetrodon* sails have revealed unexpected insights into its soft-tissue composition. Each bony spar supporting the sail has variable bone surface texture, implying that three types of tissue covered the sail in life. The base was anchored in muscle (as expected—the struts composing the *Dimetrodon* sail are the same structures that support back muscles in all vertebrates), and much of their length was encased in skin webbing. The exact nature of this tissue is unknown, but healed fractures on sail spars show it was strong enough to hold spines in place when they snapped. A third texture type is found at the tips, implying that the spines projected above the tissues of the sail webbing. Some *Dimetrodon* spines twist into unusual directions at their ends, perhaps further evidence of their liberation from the webbing beneath them.

Helicoprion *(Permian)*

THE WORD *ENIGMATIC* IS OVERAPPLIED TO FOSSIL ANIMALS, BUT there's little doubt that *Helicoprion*, a cartilaginous fish that swam Early Permian seas across the world, warrants such description. Approximately one hundred specimens of *Helicoprion* are known, and virtually all of them represent the same body part: a spiraling whorl lined with numerous (sometimes over 130) triangular, laterally compressed teeth along the outer margin. The cartilage jaw apparatus that housed this bizarre structure, as well as the rest of the *Helicoprion* skeleton, has long proved elusive, leaving us puzzling over its function and anatomical configuration since the first *Helicoprion* fossils were discovered in the late 1800s. Artists have done their best to make some implausible interpretations look convincing. Was it a projecting structure from the upper jaw, somewhat like a coiled swordfish bill? Did it droop from the lower jaw like a ragged, spiraling beard of teeth? Was it located in the mouth somehow? Was it even a jaw—might it have been a bizarre fin?

The recent discovery of a *Helicoprion* specimen with preserved jaw tissues has brought this century-long mystery to an end. It seems the whorl was, in fact, the majority of the lower jaw, with the largest teeth being the "active" teeth in the mouth. Older, smaller teeth and whorl elements spiraled into the jaw as the animal grew. The whorl is thus somewhat comparable to a shark jaw where new teeth rotate into place as old ones are pushed out, with the distinction that *Helicoprion* maintained its entire dental history within its jaw rather than shedding teeth as they left active use. The same *Helicoprion* specimen that revealed this surprising anatomy also suggests that the upper jaw lacked dentition entirely, and that the lower jaw only possessed a single row of teeth: a configuration recalling a circular saw blade and its guard more than a conventional oral cavity. It seems that only the front and top of the tooth whorl were exposed, and that as the mouth opened and closed the lower jaw moved forward and backward. This mechanism, in concert with the recurved teeth, may have acted to snag and cut prey while also drawing it into the mouth. A lack of wear on *Helicoprion* teeth suggests that they ate largely soft-bodied animals like cephalopods (octopus, squid, and their shelled relatives, such as nautiloids, shown here). The tooth whorl was prohibited from biting the upper jaw by a strap of cartilage that stabilized the side of the lower jaw and limited how far the mouth could close.

Unfortunately, while the mystery of *Helicoprion* jaws is on the way to being solved, many questions about the rest of the animal remain unanswered. It seems that *Helicoprion* belongs to a fish clade known as the Eugeneodontida, a lineage related to sharks and rays, and that its closest living relatives are the chimeras, a large group of small-to-medium-sized, sharklike fish. But we have no concrete ideas about the anatomy of *Helicoprion* beyond its jaws: even basic information, like body size and general proportions, is unknown. This limits our understanding of its lifestyle as well as its appearance, and any full-body restoration you see of this fish is—at best—an educated guess. We can, however, be sure that *Helicoprion* was a very successful and widespread animal. Fossils of *Helicoprion* are found in rocks all over the globe, and sometimes in relative abundance. It seems to have endured for at least fifteen million years and diversified into several species in that time. A solitary record from the Triassic Period has been argued as evidence that *Helicoprion* survived into the Mesozoic Era, but modern researchers are skeptical of this interpretation because the source of this fossil was poorly documented by its discoverers. Associated geological details of the alleged Triassic *Helicoprion* specimen match those of other Permian examples and, in all likelihood, its interpretation as a Triassic specimen was erroneous.

Dicynodonts (Permian)

THE ADAPTIVE POTENTIAL OF SYNAPSIDS IS WELL DEMONSTRAT-ed by the dicynodonts, a widespread, abundant, and diverse group of herbivores that existed across much of the planet during the Permian and Triassic Periods. Dicynodonts developed a more complex approach to herbivory than the caseids we met earlier, being able to slice plant matter in their jaws before digesting it in their large guts. Their efficient herbivory may explain how they became the dominant herbivorous species for much of the Permian and Triassic, before meeting their end at the close of the latter period. A fragmentary jawbone from Cretaceous rocks in Australia has hinted at dicynodonts surviving over one hundred million years beyond their assumed extinction, but this exciting claim has now been compelling challenged, and is likely incorrect. Instead, the specimen probably represents a badly preserved jawbone from a Pliocene or Pleistocene marsupial mammal.

Dicynodonts evolved in the Middle Permian from the therapsid branch of Synapsida. Therapsids showed greater mammal-like qualities than synapsids like *Dimetrodon* in having more upright limbs; hands and feet that were losing their lizard-like asymmetry; mammal-like skulls, teeth and sensory capabilities; and they seemingly possessed elevated growth rates and body temperatures. But while these were important steps toward defining mammalian traits, we should be careful not to overstate their "mammaliness." Early therapsids like dicynodonts still had some way to go before attaining truly mammal-like anatomy, and we have yet to find evidence of characteristic mammalian soft tissues, like fur, in these species.

The rapid spread of dicynodonts made them the most abundant large-bodied terrestrial animals on Earth by the end of the Permian. Some genera, like *Lystrosaurus*, were so widespread that they occurred on multiple continents, and their fossils were instrumental in validating early ideas about continental drift. In form, dicynodonts ranged from small, long-bodied species that were adapted for burrowing (including *Cistecephalus microrhinus*, the small animal in the right of this image), to elephant-sized, 4.5 meter long and 5-7 tonne megaherbivores. All were robustly built with proportionally large heads that generally lacked teeth except for, in most species, prominent tusks (or tusklike structures, in some unusual Triassic forms). They used narrow, deep beaks to crop plant matter, an action aided by powerful jaw muscles. Their beaked lower jaws slid backward when closed to create a shearing motion that chopped food in a scissorlike fashion. Chewing or slicing food before swallowing is a great advantage for an herbivore in that it helps break down tough plant tissues before digestion, leading to readier absorption of their nutrients. Chewing mechanisms have evolved numerous times among animals for this reason, and we often see plant chewers, rather than gulpers, as dominating herbivore ecological space throughout geological time. Many dicynodonts have bosses and other ornamental structures on their faces (including the South African Permian species, *Aulacephalodon peavoti*, shown here), as well as reinforced skull bones that—along with their tusks—may indicate antagonistic or defensive behaviors. They were likely common prey for predatory synapsids that existed in the Permian (such as the famous saber-toothed gorgonopsians) and for the various types of carnivorous reptiles that lived in the Triassic.

Dicynodonts were among the few major components of Permian synapsid ecosystems to survive into the reptile-dominated world of the Mesozoic, where they continued to be the major land herbivores until dinosaurian plant eaters took over their roles. Many other Paleozoic lineages were not so fortunate, however: for most species alive at the end the Permian, events were about to take a very, very bad turn.

The Great Dying: The End-Permian Extinction

AS THE PERMIAN PERIOD DREW TO A CLOSE, LIFE ON EARTH WAS subjected to its most grueling test: an extinction event that wiped out around 95 percent of marine species and over 70 percent of tetrapods. Evolutionary history was essentially reset as major lineages became extinct and biological communities crumbled, collapsing much of the ecological complexity attained in the Paleozoic and wreaking havoc in terrestrial and aquatic ecosystems alike. No other extinction event of the last few hundred million years had such an impact on life as the end-Permian catastrophe, or, as it is often called, "the Great Dying."

Some controversy exists over the exact cause of the Permian crisis. The late Permian was already a somewhat turbulent period for life because the ever-wandering continental plates had drifted into one another, forming the supercontinent Pangaea. Supercontinents have formed and split apart multiple times in our planet's 4.5-billion-year history, but Pangaea is the only supercontinent to have existed during the Phanerozoic Eon. The formation of a supercontinent has a number of climatic and environmental effects, including the reduction of coastline and shallow marine habitats. These are among the most speciose environments on the planet, so their removal leads to a rapid reduction in diversity, and we see this reflected in latest Permian fossil beds. This in itself does not explain the Great Dying, however, because the dramatic loss of life at the end of the Permian is very sharply observed and not the result of a long decline as continents slowly coalesced into one landmass. Moreover, while the construction of Pangaea led to loss of life in some communities, other environments responded positively to its formation, achieving greater diversity. It seems more likely that the cause of the end-Permian extinction was a relatively sudden event 252 million years ago, at the very end of the Paleozoic Era.

A major geological episode precisely dating to this interval was probably the major catalyst for the Permian extinction event. At the close of the Permian, a tremendous eruptive phase began in what is now northern Siberia, launching enormous amounts of ash and greenhouse gases into the atmosphere. The scale of this event is unparalleled in the Phanerozoic. Geological data suggests that the eruption spewed lava and other pyroclastic material over two million square kilometers of northern Russia. This output stemmed not from stereotyped prehistoric volcanoes—the conical sort you see in the background of unimaginative paleoart—but from fissure eruptions, splits in Earth's crust that can run for many kilometers and release massive amounts of volcanic material. The Permian fissure eruptions persisted for hundreds of thousands of years, dimming the planet as they pumped ash and dust into the atmosphere, causing acidic rain, and raising global temperatures through greenhouse gases. The fact the eruptions occurred through coal beds worsened this effect considerably, as it liberated huge volumes of carbon dioxide—a potent greenhouse gas—into the atmosphere.

Marine life was hit especially hard during the Permian event. Skyrocketing carbon dioxide levels raised oceanic acidity, making it difficult to create calcareous shells and skeletons, and temperature shifts altered oceanic circulation. Marine oxygen levels dropped and surface water temperatures rose to intolerable levels—perhaps as high as 40°C. Life on land also endured extremely high temperatures, turning lush, diverse forested habitats into barren vistas. For several hundred thousand years, life was baked and choked to death, and came the closest it ever has to total annihilation.

The Foundation of the Modern Age (Triassic)

MOST OF OUR INTEREST IN MASS EXTINCTIONS IS TUNED TO THE exciting parts: the cataclysmic events, the extreme environmental conditions, the huge taxonomic death tolls. But the aftermaths of these events are just as important in our consideration of life's evolution. The effects of mass extinctions are not easily shrugged off, often leading to cascading, runaway impacts on environmental conditions and global habitability lasting millions of years after the initial extinction episode. Our biosphere is a robust and adaptable one, but if basic foundations of climate, seawater chemistry, and oceanic circulation are knocked too far from their typical operation, the return to habitable conditions can happen only at a geological pace. These findings are arguably some of the most important to stem from paleontological research and are especially significant in light of growing concerns about the rapidly changing climates, biodiversity crisis, and shifting oceanic conditions of our modern age.

The aftermath of the end-Permian extinction was a difficult period for life in both terrestrial and marine settings. The absence of extensive vegetation on land reduced the diversity of habitats and the availability of food, as well as exposed soils and bedrock to erosive forces, thus allowing large volumes of terrestrial debris to be swept into bodies of water. This may not seem too significant—what harm can a little wind-blown sand do?—but it actually created chaos in the oceans. High sediment influxes compromised the ability of marine species to feed, reproduce, regulate their internal salt balances, and develop strong skeletons. They promoted blooms of microbes that, upon death, had a stagnating effect on oceanic oxygen levels and further stressed marine life. Some aquatic habitats were so inundated with sediment that species adapted to living in or on soft substrates thrived, but those adapted for other habitats went into decline. With global temperatures and oceanic acidity also remaining generally high, it's unsurprising that many marine species—even those that survived the initial end-Permian event—did not endure these conditions for long. About five million years passed before biodiversity started to recover, and it took another few tens of millions of years to achieve pre-extinction diversity levels.

The cast of animals that repopulated Earth in the Mesozoic Era was not the same as that of the Permian. Trilobites, eurypterids, acanthodians, several major coral and insect groups, numerous synapsids, and some early reptile lineages were gone. Animals that anchored to the seafloor—such as the once-abundant brachiopods—were especially hard hit, and they never recovered their former diversity. What replaced them were animal types and ecosystems that we would find more recognizable: the basis of our "modern" natural world was born out of the ashes of the Permian. New types of corals, bivalves, snails, and sea urchins became abundant in marine environments and, on land, ecosystems were populated by the mammal-like cynodonts and a hitherto fairly innocuous evolutionary lineage: the reptiles. With reptiles having largely been sidelined by synapsids since their evolution in the Carboniferous, the Triassic marked the interval where they made a play for world domination. Starting their time in the Triassic as relatively unassuming, somewhat crocodile-like species—such as the Brazilian *Teyujagua paradoxa* (shown here, middle right) or its squat contemporary *Procolophon trigoniceps* (bottom)—by the end of the period reptiles had come to dominate land ecosystems, had achieved a tremendous diversity of seagoing forms, and had even adapted to powered flight—a first for vertebrates. Dicynodonts showed some resilience against the rise of reptiles but relinquished control of herbivorous niches to saurian lineages at the end of the Triassic. This surge in reptile evolution saw several major groups competing for roles as dominant terrestrial predators and herbivores as the Triassic drew to a close: a highly adaptable and successful clade, the dinosaurs, were the principle victors.

The Basilisk Lizard, a "Lower Vertebrate" (Holocene)

THE REPTILES WE KNOW TODAY GIVE ONLY A NARROW, AND PERhaps somewhat misleading, characterization of this group compared to their anatomical and ecological diversity throughout geological history. Modern reptiles are the turtles, the crocodylians, and the lepidosaurs (the group that includes lizards, snakes, and tuatara). They are easy to classify from their scaly bodies, sprawling limbs, and environment-controlled "cold-blooded" physiologies (all features seen in the common basilisk, *Basiliscus basiliscus*, right). If we consider the broader picture of reptile evolution, however, we find that it's much harder to stereotype what a reptile is. In addition to species like those alive today, the reptile fossil record contains "warm-blooded" animals that had elevated metabolic rates, upright postures, and fuzzy skin. Ancient reptiles include dinosaurs, pterosaurs, several great marine lineages, and a number of bizarre forms that defy easy categorization—all animals with lifestyles and anatomies very different from those of their modern cousins. We also have to classify birds as a type of reptile, their ancestry being among the predatory dinosaurs. From this perspective, reptiles are not just the scaly sprawlers we know today, but an evolutionary tour de force, a lineage that has been prominent in a range of habitats across the world since the Triassic.

Reptile phylogeny has been studied intensively and the broad picture of their evolution is now well understood, though the origin of turtles remains challenging to pin down (we'll return to this later). Lepidosaurs represent one major branch of the reptile line, the other belonging to the archosaurs and their relatives. Archosaurs are the evolutionary home of crocodylians and their kin, as well as the lineage that gave us pterosaurs, dinosaurs, and birds. The origins of these groups are ancient, with lepidosaurs and archosaurs first appearing in the Permian, and the archosaurs splitting into the bird and crocodylian branches in the Triassic. With such deeply rooted divisions, it's not surprising that living reptiles are only superficially similar. Once we look closely, we find they are quite distinct from one another in detailed anatomy.

Humans have often regarded reptiles as second-class animals despite their evolutionary success. Scientists have historically considered reptiles, along with amphibians and fish, as "lower" vertebrates: dim-witted, anatomically primitive animals that are physiological and behavioral shadows of "higher" species, the birds and mammals. In broader culture, reptilian features—scales, "cold-bloodedness," and crawling gaits—are often applied to subjects of revile or suspicion. But ongoing studies of the "lower" vertebrates demonstrate that our traditional view of their physical abilities and intelligence has been biased by the idea that humans, and species most biologically similar to us, are an evolutionary peak that all organisms aspire to. For instance, while upright limbs may confer energetic advantages over sprawling configurations, sprawlers enjoy greater stability and advantages in climbing, sprinting, and frequent acceleration. These properties strongly suit the lifestyles of many small tetrapods, and their retention of this so-called "primitive" stance likely reflects the fact that a mammal-like upright limb posture would compromise their evolutionary fitness. Similarly, a reptilian physiology requires only about 10 percent of the food needed by an equivalently sized warm-blooded creature, making reptiles better suited to life in areas of low productivity than a mammal or bird. The idea that a slow metabolism is associated with lack of behavioral sophistication or low intelligence has also not been borne out under study. Reptiles seem to have good memory, flexible (instead of purely instinctive) behavior, and problem-solving abilities. They even engage in play behavior, just like mammals and birds. Crocodylians will play with tethered balls, and turtles engage in tug-of-war with zookeepers or toss objects gleefully around their ponds. Komodo dragons, the largest living lizards, take part in especially sophisticated recreational behavior where they play with their handlers and make toys of balls, buckets, and old shoes. These so-called lower vertebrates are far from second-fiddle to mammals and birds: they are our adaptive peers, and their prominence and persistence in Earth's evolutionary history is far from a fluke.

Amphibious Ichthyosaurs (Triassic)

IT'S A SOMEWHAT PERVERSE FACT OF EVOLUTIONARY HISTORY that, within a few million years of amniotes evolving to live independently of water, several lineages performed an adaptive U-turn to reinvade the aquatic realm. Of course, by this point amniotes had long abandoned key adaptations to aquatic existence: they had limbs instead of fins, they breathed air, and they laid eggs that would drown if submerged. If we ever need an example of how random, opportunistic, and counterintuitive natural selection can be, the evolution of secondarily aquatic Permian and Triassic amniotes is a great example.

The first of these U-turns was launched by the mesosaurids, a group of lizard-like Permian animals with uncertain relationships to reptile-line amniotes. Mesosaurids seem to have been a relatively short-lived experiment with secondary aquatic lifestyles however, and it was not until the Triassic that amniotes fully committed to life at sea. From the Triassic through to the end of the Cretaceous, Earth's seas were full of reptiles: ichthyosaurs, plesiosaurs, mosasaurs, turtles, and many other unusual forms of uncertain evolutionary affinities. Many of these reptiles were as comfortable in the water as a whale or a dolphin, being efficient swimmers; accomplished aquatic predators; and capable of giving birth to live offspring, negating the need to leave the water to lay eggs. Making this more remarkable is the fact that "marine reptiles" are not a single group but several different lineages that entered the seas independently, converging on the same adaptations for wholly marine lifestyles. Quite how many times reptiles developed aquatic adaptations is a matter of contention, as the evolutionary origins of some groups—including the famous ichthyosaurs and plesiosaurs—remain difficult to pin down.

The Triassic was a special time in marine reptile evolutionary history, thanks to the presence of many unusual lineages that would not survive into the later Mesozoic. Among them was *Cartorhynchus lenticarpus*, a small (about 40 cm long) Chinese marine reptile that represents an early stage of ichthyosaur evolution. The ichthyosaurs modified their bodies for swimming more than any other aquatic reptile, eventually assuming a fishlike form entirely unlike that of their terrestrial ancestors (hence the name "ichthyosaur", which translates to "fish lizard). But they did not attain this anatomy overnight. The earliest ichthyosaurs retained several relatively "terrestrial" characteristics compared to later forms, including a long, slender tail instead of a tail fin, relatively long hind limbs, short bodies, and generalized skulls and teeth. Many of these features were present in *Cartorhynchus*, as were unusually long, flexible and curving forelimbs that may have allowed this species to move about on land. *Cartorhynchus* was clearly a primarily aquatic animal however, possessing dense, heavy bones to reduce its buoyancy, true flippers instead of walking limbs, and a jaw apparatus adapted to suction feeding (an aquatic foraging mechanism where prey is sucked into the mouth though quick, powerful expansion of the mouth cavity). If *Cartorhynchus* did amble about on land, it probably was not especially speedy or agile. Our mental image might be better informed by sea turtles and mudskippers than by seals or otters. Perhaps it left water only to escape predators, to rest, or to access water bodies that were cut off from the sea.

Curiously, though *Cartorhynchus* seems like a "transitional" species, moving from life on land to life in the oceans, its placement in current schemes of ichthyosaur evolution suggests that it arose from species that were already strongly adapted to marine life. Thus, if *Cartorhynchus* was an amphibious ichthyosaur, it represents an aquatic species adapting to life on land, not the other way around. Perhaps *Cartorhynchus* represents an ichthyosaur lineage that never committed to land or sea, but instead ebbed between the two as adaptive opportunities arose.

Rebellatrix *(Triassic)*

THE TERM *LIVING FOSSIL* HAS A NUMBER OF DEFINITIONS. ONE applies to modern animals that closely resemble their fossil relatives, where slow evolutionary rates preserve anatomy developed by ancient, now-extinct relatives until modern times. A second definition applies to species known from fossils before living examples of the same lineage are found alive in the modern world. Such species are sometimes called "Lazarus taxa"—organisms that have, like the biblical Lazarus, come back from the dead.

Both definitions are met by the coelacanth (*Latimeria*), a large (2 m long) and robust lobe-finned fish most closely related to lungfishes and tetrapods. Coelacanths have a good fossil record from the Devonian to the Cretaceous as well as a long research history, their fossils first being studied in the early 1800s. 19th century paleontologists found that the coelacanth fossil record ran dry in rocks around seventy million years old, so they sensibly assumed that coelacanths were long extinct, perhaps victims of the same mass extinction that wiped out many marine species at the end of the Cretaceous Period. We can only imagine the surprise paleontologists felt when a living coelacanth was caught in the Indian Ocean in 1938, and especially because this living example was superficially little different to those known from the geological record—a true "living fossil" in every sense. Two living species of coelacanth are now recognized: the critically endangered West Indian coelacanth (*Latimeria chalumnae*) and the Indonesian coelacanth (*Latimeria menadoensis*).

Modern fish experts are pushing back against the concept of coelacanths being "living fossils", however. First, coelacanth DNA shows a typical rate of mutation and change, without any evidence of slow or static mutation rates. This is not unexpected, because although the deep-water habitats occupied by coelacanths are often regarded as archaic "lost worlds," they are actually just as dynamic and changing as any other setting. There is no reason why coelacanths should have, or would benefit from, reduced evolutionary rates. Second, fossil coelacanths were very diverse, with over 130 known species in a myriad of sizes and forms. Variations in body proportions represent vastly different approaches to locomotion, and differences in jaw anatomy hint at different dietary preferences. Some coelacanths were huge, 5-m-long ocean cruisers, while others were small, compact fish better adapted to skirting over the seafloor than living in the open sea. We know that many ancient coelacanths lived in freshwater habitats instead of marine settings, and that swim bladder (organs that help fish regulate buoyancy) anatomy differs fundamentally between living and extinct species, suggesting different habitual water depths. Just one example of a fossil coelacanth unlike *Latimeria* is the 1.3-m-long Triassic species shown here, *Rebellatrix divaricerca*. The most obvious feature of this Canadian coelacanth is its large forked tail, a fin morphology unknown in other coelacanths and thought to be linked with a powerful, sustained swimming ability. The skull of *Rebellatrix* is unknown so its exact diet is not clear, but its streamlined body shape and powerful tail compare well to certain sharks and predatory fish, suggesting that it may have been a fast pursuit predator. This lifestyle is very different from the relatively sedate bottom-feeding habits of living coelacanths, which exploit deep-sea currents to drift over foraging areas, swimming only when necessary.

It is thus difficult to consider modern coelacanths "living fossils." There is no stereotyped coelacanth anatomy or lifestyle for them to conform to, just as there is nothing sluggish about their rate of evolutionary change. Indeed, the whole concept of "living fossils" is a problematic one: any study of life that delves beneath superficial details invariably shows that evolution and adaptation are rarely static.

Erythrosuchids (Triassic)

MANY TRIASSIC REPTILES HAVE A CHARMING CRUDENESS ABOUT their appearance, where fairly conventional reptilian bodies and limbs are married to a distinctive, often bizarre head and neck anatomy. This can appear as if evolution was somehow rushed, failing to adapt the entirety of these animals to specific lifestyles and instead just changing a few bare essentials. There may be a nugget of truth to this. Many Triassic animals were rapidly filling ecosystems emptied by the Permian mass extinction, and without significant competition from established species, they may not have needed to optimize their anatomy as far as later occupants of the same niches. But appearances can be deceiving, and we should not assume that these animals were somehow "inferior" to later forms. Many of these Triassic reptiles, despite their seemingly "crude" appearances, belonged to long-lived and widespread lineages and held their own against more sophisticated-looking reptiles of the later Triassic.

Among these unusually proportioned species were the erythrosuchids. These giant-headed reptiles enjoyed twelve million years of evolutionary history through the Early and Middle Triassic world, their fossils remaining undiscovered only in North America and Antarctica. They are most closely related to the archosaurs, though they lack close affinity with any living reptile group. Many erythrosuchids are known from relatively complete skeletons, and we have a good idea of their overall proportions and skeletal anatomy. Some species were large, reaching nearly 5 m in length, though others—like *Garjainia madiba*, shown here—were just over 2 m long. Their most characteristic feature is their enormous heads and jaws, the function of which is betrayed by their sharp, recurved and predatory teeth. Erythrosuchid heads were probably not as encumbering to their owners as they look. Erythrosuchid snouts were air-filled and narrow, so only the back of the head—which housed massive muscles to power their mighty jaws—was

especially broad and heavy. Thus, their heads were probably lighter than they appear, and we can read their anatomical indications of expanded neck muscles and powerfully built forelimbs as mechanisms to wield their oversize jaws with fine control. What we know of their limbs suggests that these were long and robust, and erythrosuchids may have been capable of semierect postures instead of a lizard-like sprawl.

Exactly how these remarkable animals made a living remains an area of investigation. Historically, the large heads of erythrosuchids were thought to dictate aquatic habits, it being reasoned that water was essential to supporting their front-heavy anatomy. In this idea, their heavy, thick-walled limb bones would help them sink and submerge in swamps and rivers. But more recent research is favoring terrestrial habits, noting features of weight reduction in their skulls, pointing out their lack of swimming or wading adaptations, and explaining their heavy limb bones as being reinforced for carrying large bodies over land. Curiously, erythrosuchids are often the largest animals of their respective environments, an unusual occurrence among purely terrestrial predators. Might this indicate that they hunted aquatic prey, at least on occasion? More research is needed to clarify the roles of these animals in Triassic ecosystems.

One of the most interesting attributes of erythrosuchids is their growth rates. They grew much faster than living lizards or crocodylians, and at a speed more comparable to many dinosaurs, early birds, and the flying pterosaurs. These elevated growth rates are thought to reflect a fast, "warm-blooded" metabolism and are typically associated with a number of other anatomies: fully upright stances, insulating fur or feathers, and large brains. But the rapidity of growth in erythrosuchids indicates that fast growth appeared much earlier in reptile evolution, and before the development of those related features.

Atopodentatus, *an Early Underwater Herbivore (Triassic)*

THE TRIASSIC IS WELL KNOWN FOR ITS PECULIAR MARINE REPtiles, but few rival the strangeness of the recently discovered *Atopodentatus unicus*. This species is represented by skeletons from Middle Triassic rocks of China that represent almost all of its bony anatomy. Despite this relative wealth of data, the relationships of *Atopodentatus* to other marine reptiles are not well understood. Provisional analyses suggest affinities with the sauropterygians, a major clade of marine reptiles that includes the nothosaurs, placodonts, and plesiosaurs. However, general uncertainty about the relationships of marine reptiles to one another, and the relatively recent nature of the discovery of *Atopodentatus*, mean further work is needed to confirm this hypothesis.

Atopodentatus is a genuinely bizarre animal, an evolutionary collision of seal, crocodile, and vacuum cleaner accessory. At 2.75 m long, it was a moderately sized marine reptile, but its skull was just 12 cm long—less than 5 percent of its body length. This was mounted atop a relatively long neck; a long, deep body; and a powerfully muscled tail. Unlike some marine reptiles, its tail skeleton lacks obvious hallmarks of a fin or a paddle, and *Atopodentatus* may have relied on its limbs, as well as its tail, to propel itself through water. Both sets of *Atopodentatus* limbs were stout and paddlelike, with long, broad fingers and toes. However, its limbs were also capable of flexing at their knees and elbows, and the pelvis was strongly attached to the spine. These features indicate *Atopodentatus* was probably capable of walking on land, though its weak wrists and ankles indicate a limited capacity to support its weight out of water. Perhaps it spent much of its time swimming, and left water only to rest, to escape danger, or—if it was incapable of giving birth at sea—to lay eggs.

The most unusual component of *Atopodentatus* is its tiny skull. Its jaws were lined with hundreds of tiny, peg-like teeth that were, when *Atopodentatus* was first discovered, thought to line the front of a downturned face in two vertical rows: imagine an unhappy reptile with a zipper on its face and you're not far off initial interpretations of this species. Subsequent discoveries showed that this bizarre interpretation was based on a broken skull, and that the jaws actually formed a T-shaped muzzle with a wide set of dental "combs" along the front margins. Further teeth lined the cheek region, and the upper internal surface of the mouth was covered with minute denticles. Its bite—judging by the available space for jaw musculature—was relatively weak, although the musculature associated with opening the jaws was expanded, and the lower jaw was robust. This peculiar arrangement of features is thought to represent a sophisticated mechanism for aquatic herbivory, where the wide dental "combs" scraped or sheared algae from underwater surfaces, and powerful opening motions of the jaws sucked dislodged food into the mouth. The plant matter was then separated from the water by the internal denticles and cheek teeth, before the sieved water was expelled from the side of the mouth.

Herbivory is relatively rare among marine tetrapods. Perhaps the only other known Mesozoic marine reptile to live exclusively off plant matter is the placodont *Henodus*, a turtle-like creature that cropped and filtered algae using unusual dentition and baleen-like structures along the side of its jaws. Today, marine iguanas (*Amblyrhynchus cristatus*) and certain sea turtles graze on algae during underwater dives, while certain mammals, such as sirenians (manatees and dugongs), live off algae and marine grasses. Fish have had far greater success as aquatic herbivores, with Cenozoic groups like parrotfishes and surgeonfish developing means not only to remove algae from underwater surfaces but also to extract microscopic plants living within rocky sediments. This has shaped marine ecosystems in limiting algal growth and facilitating development of grazing-resistant communities, including modern-grade coral reefs. Evidence of these habitats first appears in Miocene sediments, alongside fossils of these remarkable fishes.

Morganucodon *and the Dawn of Mammals (Triassic)*

OF THE SYNAPSIDS THAT SURVIVED THE PERMIAN EXTINCTION event, only the dicynodonts and a group yielding our own ancestors, the cynodonts, conducted substantial evolution in the Triassic and beyond. The cynodonts, a group originating in the Permian, possessed many features we associate with true mammals: a heightened metabolism, the start of the transformation of the posterior lower jawbones into the inner skeleton of the mammalian ear, and—in some later groups—the probable development of fur and whiskers. Direct evidence for the earliest appearance of fur remains elusive, but sensory pathways in some cynodont skulls match those of whiskered mammals more than those of hairless, nonmammalian species. If correctly interpreted, this implies the development of sensitive facial tissues and whiskers in some particularly mammal-like cynodonts. Our understanding of cynodont evolution, and the discovery of hairlike structures in a coprolite (fossil excreta) indicate that this development probably took place in the late Permian.

Morganucodon watsoni represents a stepping stone between cynodonts and the true mammals. Technically speaking, *Morganucodon* is a mammaliaform rather than a true mammal: a very mammal-like creature but not part of, or closely related to, any living mammal group. Numerous *Morganucodon* bones have been found in ancient fissure fill deposits in Triassic/Jurassic rocks of today's United Kingdom, allowing experts to reconstruct aspects of its skeleton from numerous individuals. *Morganucodon* was the first mammaliaform known from such extensive fossil material and was a very welcome discovery given that, until it was found in the mid-twentieth century, paleontologists knew Mesozoic Mammaliaformes largely from isolated bones and teeth.

With a body length of 10–13 cm (excluding the tail), *Morganucodon* would have resembled a small rodent in life, though its bowed limbs would contrast with the rats and shrews it is often compared to. Its skull was stoutly built, with expanded cavities to anchor jaw muscles. These, and its sharp, piercing teeth, imply a diet of hard-shelled invertebrates and other small animals. Its ears had yet to make the full migration from the lower jaw to the skull, so they would have been set lower on the head than we're accustomed to in living mammals. It is not yet clear whether *Morganucodon* had ear pinnae—those conspicuous skin and cartilage structures that characterize most modern mammalian ears. Monotremes (the echidna and platypus, egg-laying mammals which represent the most ancient grade of mammal evolution to survive to modern times) have very small pinnae or none at all, and exceptionally preserved fossils of the aquatic Cretaceous mammal *Castorocauda lutrasimilis* also lack conspicuous ear tissues. Might these examples suggest that, as shown in this illustration, *Morganucodon* lacked ear pinnae? Possibly, but we need more than three lines of anatomical evidence to know for certain.

Scurrying away with a kidnapped *Morganucodon* puggle is a small lizard-like species known as *Clevosaurus bairdi*. This is not a lizard, however, but a sphenodontian: a relative of the modern tuatara. Sphenodontians and lizards share ancestry on the same branch of the reptile tree, and both arose in the Triassic. Sphenodontians were the more successful group to begin with, however, spreading all over the world and diversifying into a number of lifestyles. It was only in the Jurassic that lizards started to attain their modern level of distribution and diversity. The once-mighty Sphenodontia survives today only as the tuatara, all species of which are found in New Zealand, while lizards occur on all continents except Antarctica.

The Great Crinoid Barges (Jurassic)

HUGE MONSTERS OF WRITHING, FLOWER-HEADED TENTACLES once roamed Jurassic seas. Cruising the waves like Lovecraftian barges, they hung ropelike structures—some over 20 m long—with 80-cm-wide dragnets sweeping the sea clean of organic matter. After years of drifting, the ever-increasing mass of these ropelike creatures surpassed the floatation capability of their vessel and they sank to the seafloor, starving to death in the still, deep sea, or else suffocating in oxygen-free bottom waters. On occasion, these behemoths would be preserved as fossils, the tendrils of some Early Jurassic German examples extending over an area of five hundred square meters. Their remains are, without doubt, some of the most spectacular fossils in the world.

Closer inspection of these fossils shows that they were not single, multitentacled organisms but collectives of several species. At their core were large pieces of driftwood, mostly pieces of tree trunks that are, in the biggest specimens, over 10 meters in length. Driftwood in the modern day can float for several years, but it is often sunk through the actions of burrowing mollusks. Such organisms had yet to evolve in the Jurassic, however, so driftwood may have remained buoyant for far longer. As the logs floated, they accumulated bivalves (clams) across much of their surface, between which anchored gigantic crinoids (sea lilies). Crinoids are a type of echinoderm, the same group that includes starfishes and sea urchins. Echinoderms are characterized by fivefold body symmetry; skeletons made out of numerous calcareous plates, spines, and segments; and the presence of tiny "tube feet"—minute structures, projecting from skeletal pores, that are used in walking and feeding.

Crinoids are among the most ancient of the echinoderm groups. Their fossil record dates back to at least the Ordovician Period, and a Cambrian origin is predicted by some scholars. They can be grouped into two major morphs: stalked crinoids, which position their flowerlike feeding cones in the water column by standing on the seafloor or anchoring themselves to submerged objects; and free-swimming species that lack stalks and move through water by beating their arms. Today the free-swimming species dominate crinoid diversity, but the stalked variants were much more common in Paleozoic and Mesozoic seas. Indeed, fossils of stalked crinoids are so abundant in Carboniferous marine sediments that their skeletons form thick limestone deposits. Paleozoic crinoids were particularly diverse in both taxonomy and lifestyle, but both attributes were reduced significantly during the Permian mass extinction. Stalked crinoids bounced back from this catastrophe in the Mesozoic and are common fossils from this time, but they never regained a diversity akin to their Paleozoic heyday.

The giant crinoids that anchored themselves to Jurassic driftwood, known as *Seirocrinus subangularis*, attached themselves as larvae and grew to huge lengths. How quickly they grew remains unknown, but it is suspected that the biggest stalks took many years to attain their great size. As free space on the driftwood became harder to find, smaller crinoids would simply anchor themselves to their larger brethren—stalks growing on stalks. Crinoids are filter-feeders reliant on water currents to bring food particles into reach of their arms, and some living stalked species walk across the seabed looking for suitable foraging areas, crawling along using short tentacles emerging from their stems. Seafaring species like *Seirocrinus* could feed continuously, however, thanks to the motion of the driftwood trawling their feeding apparatus through the water. With relatively little nutritious soft tissue in their bodies, these giant crinoids might have been relatively ignored by predators, but their lives were not without danger. Some *Seirocrinus* fossils show that they lost their feeding apparatus entirely, their dead, headless stalks trailing in the sea as the barges sailed on.

Ophthalmosaurus *(Jurassic)*

OUR PREVIOUS MEETING WITH A MEMBER OF THE ICHTHYOSAUR lineage featured a Triassic species still bearing features of terrestrial animals, and thus someway off deserving the title of a true "fish lizard." Later ichthyosaurs, like the European Middle Jurassic species *Ophthalmosaurus icenicus*, had fully earned this name, however. Advanced ichthyosaur features included a long torso with a deep chest, limbs transformed into stiffened flippers (larger at the front, smaller at the back), a large tail fin, and a soft-tissue dorsal fin on their backs. This last feature is demonstrated by well-preserved Jurassic ichthyosaur fossils, which record soft-tissue body outlines as dark stains surrounding the skeleton. The nature and reliability of these stains has been historically controversial: are they the remnants of microbes that once covered the tissues of the decaying animal, or genuine ichthyosaur bodies? The authenticity of the preserved outlines has also been questioned, in that components like dorsal fins have been interpreted as wayward body tissues dislodged during decay. Such concerns were further complicated by the tendency of nineteenth-century fossil preparators to "clean up" specimens, removing untidy edges or even plastering over anatomical details to make the fossils more aesthetically pleasing. This practice was not limited to ichthyosaur fossils: many specimens collected in the early days of paleontology were "improved" for aesthetic reasons, often to the later detriment of those trying to interpret a specimen's real anatomy.

Continued study of ichthyosaur fossils has demonstrated that their dorsal fins and body outlines are genuine records of their tissues, and they provide us with a lot of data about their life appearance and hydrodynamics. Alas, such exceptional remains are only associated with a handful of species, so the nature of fin and body shapes across ichthyosaur evolution remains poorly understood. Given the varied skeletal proportions and body sizes of ichthyosaurs, some diversity in soft-tissue form might also be expected. Some Triassic ichthyosaurs were relatively eellike, with low tail fins and flexible bodies. Later forms, including *Ophthalmosaurus*, attained a "thunniform" (tunalike) body plan that permitted particularly efficient swimming. Like modern whales and fast-swimming fish, only the end of the thunniform ichthyosaur tail was flexible, an adaptation to maximize thrust generated by the tail while swimming. The tail fin itself was tall and crescent-shaped, and thus capable of generating considerable forward momentum. The rest of the body was deepened and broadened, achieving a body form optimized for aquatic locomotion.

Collectively, these adaptations would have made ichthyosaurs fast, powerful swimmers. Their teeth and stomach contents suggest a diet of fish, squid, belemnites (relatives of squid and octopuses with substantial internal skeletons), and—in larger species—other marine reptiles. In this respect we can analogize them with living toothed whales, such as dolphins and orcas. Their commitment to marine life was so great that they had no ability to walk on land and they gave birth to live young at sea, an adaptation precluding the need to leave water to lay eggs. The occurrence of up to eleven embryos in one pregnant ichthyosaur fossil suggests they gave birth to many offspring at once. Such strategies are used among living animals with high juvenile mortality rates, implying that survival as a juvenile ichthyosaur may have been fraught. Perhaps baby ichthyosaurs relied more on the odds of their brothers and sisters being eaten than on having nurturing parents to help them survive.

Ophthalmosaurus is characterized by the largest set of eyes of almost any animal. At 23 cm across, its eyeballs were second only to those of the modern giant squid, *Architeuthis dux*. When looking at a living *Ophthalmosaurus*, however, we would have seen only a portion of their eyes, as most of the eyeball was hidden behind bone and skin. But even with only partial exposure, their eyes would have been remarkably sensitive to light, and this likely permitted *Ophthalmosaurus* to find prey in deep water or other dim conditions where other animals were rendered sightless.

Anchiornis: *A Dinosaur That Was Almost a Bird (Jurassic)*

MOST OF US ARE FAMILIAR WITH THE IDEA THAT THE EVOLUTION of birds had something to do with dinosaurs. Less well known is the exact nature of the relationship between these groups and the extensive fossil record that entirely blurs any hard boundary between them. Thousands of feathered dinosaur fossils—many of them from China—document the evolution of birds from theropod (predatory) dinosaurs, with birdlike features especially obvious in the clade that we call Paraves. Mesozoic paravians were feathered animals that include species like *Deinonychus* and *Velociraptor* among their anatomically and ecologically varied forms. For much of the Mesozoic, birds were just one type of two-legged, feathered animal among many, and it's likely that time travelers to Mesozoic settings would struggle to distinguish "true" birds from several types of nonavian dinosaur. This brings home the real fact of bird origins: birds are not *related* to dinosaurs; birds are not *something to do* with dinosaurs; *birds are dinosaurs*. It's sometimes queried whether birds "stopped" being dinosaurs once they began evolving independently from other theropods, but this cannot be so. Birds can no more stop being dinosaurs than we can stop being mammals. This means that every bird we see today—even a familiar or comical bird like a pigeon, a chicken, or a parrot—is a living dinosaur, and we can no longer consider dinosaurs to be extinct. To the contrary, nearly ten thousand dinosaur species live on Earth today—far from being extinct, dinosaurs are one of the most diverse vertebrate groups of modern times!

Many features we associate with modern birds are actually rooted in their dinosaur ancestry, and some originated even more deeply within their archosaurian origins. These include their hollow bones, their large brains, a body covered with feathers (or early versions thereof), and even anatomical minutiae like wishbones. The acquisition of avian characteristics was not a sudden burst of evolution specific to one type of dinosaur, but a gradual accumulation of features that produced many birdlike creatures. Animals that might qualify as "the first birds" appear about 165 million years ago in the Middle/Late Jurassic, though distinguishing fossils of the first "true" birds from "very birdlike dinosaurs" is increasingly difficult. Evolution is a continuum rather than a neat series of nested categories, and when our fossil record is relatively complete we can struggle to find obvious boundaries between our taxonomic groups. So blurred is the evolution of birds from nonbird dinosaurs that any distinction between them is now almost arbitrary.

What sort of dinosaurs did birds evolve from? *Anchiornis huxleyi* is a small paravian from Jurassic deposits of China and is a fairly typical "dino-bird." It is known from dozens of exceptionally preserved fossils that preserve soft tissues along with entire skeletons, some of which even preserve details of feather coloration (reflected in the illustration opposite). *Anchiornis* was a long-limbed bipedal animal with a delicate, short-snouted skull; a relatively large brain; and small, sharp teeth. It was covered—literally head to toe—in feathers of various kinds, with simple downy feathers across much of the body and vaned feathers on its arms, legs, and tail. The feathers on the limbs were especially long and give the impression of *Anchiornis* having four wings, a condition mirrored in several other paravian species. It retained a three-fingered hand, but the second and third digits were married together by soft tissue, a precursor to the bony fusions between the second and third fingers of later birds. *Anchiornis* had abandoned the heavy, muscular tail typical of reptiles for a lightweight structure composed of slender vertebrae. As a small, lightweight animal, it was well suited for rapid movement through the forests in which it lived, most likely in pursuit of a diverse, omnivorous diet. *Anchiornis* was probably incapable of flying, however, its wings being too small to sustain flight and its chest skeleton lacking adequate space for flapping muscles. Not all paravians were flightless, though: multiple lineages experimented with different wing apparatuses and flight mechanics, of which the modern avian approach (two feathered wings and flapping flight) is just one configuration.

Brontosaurus (*Jurassic*)

BRONTOSAURUS IS A NAME THAT MANY OF US ASSOCIATE WITH historic aspects of paleontology: the Golden Age of nineteenth-century dinosaur collecting, twentieth-century imagery of swampbound reptiles, and the 1903 suggestion that this most famous dinosaur was likely just a type of *Apatosaurus*. This idea saw the term "*Brontosaurus*" fall out of scientific use for over a century, although nostalgia for the name meant it was never entirely abandoned. Even Knight, in this book's 1946 predecessor, used "*Brontosaurus*" forty-three years after scientists considered the name invalid. Happily for those with a penchant for vintage dinosaurs, a careful analysis of *Brontosaurus* and *Apatosaurus* fossils published in 2015 showed that they were not quite as similar to each other as once proposed, and we can again distinguish *Brontosaurus* as a valid type of dinosaur.

Brontosaurus is a sauropod, the group of long-necked dinosaurs that includes the largest land animals of all time. The exact length and magnitude that these animals could obtain is a point of contention, as the fossils of the largest individuals are invariably incomplete, and estimating their weight is fraught with challenges. However, lengths of 30 m or more, and body masses exceeding 50 tonnes, are conservative assessments of their maximum size. Only the largest whales can best sauropods in body size, and they have the benefit of water to support their mass. As land animals, sauropods had to rely on spectacular weight-saving anatomies to become so gigantic. Chief among these were extensive air sacs in their bodies and skeletons, features that helped their tissues grow to huge proportions without adding much additional mass. Most dinosaurs bore air sacs in their bodies, but no others capitalized as much as sauropods on their utility to permit large size.

Today, we know that *Brontosaurus* is not the round-faced, swamp-dwelling animal many of us remember from our childhoods. Rather, *Brontosaurus* belongs to a generally gracile group of sauropods known as diplodocids: animals characterized by their proportionally long necks; exceptionally long tails with whiplike ends; relatively narrow bodies; and slender, horselike skulls. It would have consumed huge quantities of plant matter, using jaws lined by relatively simple, peg-like teeth. The guts of sauropods were so long that plant tissues could be broken down effectively within the stomach and intestines, thus negating any need for sauropods to chew their food. This meant that sauropod skulls didn't need robust teeth or jaws and were light enough to be perched atop long necks, giving them tremendous reach to harvest food. But despite appearances, sauropod heads are not atypically small. Investigations into animal scaling show that the skulls of land-based herbivores generally become proportionally smaller as their bodies get larger, and we are simply seeing this relationship expressed to an extreme degree in these giant reptiles. As with all diplodocids, *Brontosaurus* seems anatomically suited to rearing up on its hind limbs, perhaps to reach higher foliage or intimidate other animals.

Brontosaurus and its close relative, *Apatosaurus*, are known for their particularly massive neck vertebrae, the significance of which has long gone without explanation. Recent work has found that much of their unusual vertebral anatomy—including truly massive, looping neck ribs and the presence of prominent knobs along their underside—could be consistent with a role in combat. Might *Brontosaurus* and kin have used their necks as thunderous bludgeoning and wrestling apparatus? Neck-based combat might seem strange because our own necks are short and vulnerable structures, but modern giraffes and seals use their necks in various types of aggressive acts, both as clubs and as wrestling aids. A roughly analogous behavior may have been occurring, scaled up many times over, in the Jurassic with *Brontosaurus*. It's hard to think of a take on this famous dinosaur that is more divorced from its classic, swampbound visage than this.

Mesozoic Mammals (Cretaceous)

OUR MESOZOIC ANCESTORS, SURROUNDED AS THEY WERE BY spectacular fossil reptiles, often register as little more than a footnote on the radar of popular science. It seems the charisma of dinosaurs, marine reptiles, and pterosaurs is enough to suppress our typical instinctive interest in our own evolutionary history, of which the Mesozoic years were a major and defining part. Since the earliest days of paleontology, Mesozoic mammals have been relatively poorly known, being represented mostly by individual teeth and bones and only rarely by more complete remains. But our understanding of their diversity and ecological breadth has improved tremendously in recent decades, thanks to new fossils of not only complete skeletons but also soft-tissue features such as hair and ear cartilage, and even dietary remains. These fossils confirm long-held views that Mesozoic mammals were primarily small-bodied (the largest known Mesozoic species is about the size of a European badger), but they also show a far more ecologically diverse group than previously realized.

Undisputed members of Mammalia—as opposed to the *Morganucodon*-like animals we met previously, which are not universally accepted as "true" mammals—appeared in the Jurassic Period and quickly diverged into the major divisions of modern mammals. The egg-laying monotremes, the pouched marsupials, and the placentals (our own group, which gives birth to relatively developed live young) were thus all in existence by the Jurassic–Cretaceous boundary. By the end of the Cretaceous, these lineages had diversified further into the earliest members of the major mammal types we know today, as well as a number of lineages that existed in only the Mesozoic and the early Cenozoic.

The archetypal Mesozoic mammal was something like *Durlstodon ensomi* (here in left foreground) or *Durlstotherium newmani* (right and center foreground) in appearance: adaptable, small-bodied creatures suited to life among low vegetation and leaf litter, primarily pursuing insects and nutritious plant matter to fuel their furry, high-metabolism bodies. It's a mistake, however, to believe the common suggestion that dinosaurs and other predatory reptiles would have been too large to concern themselves with our ancestors: many smaller dinosaurs, including many paravians, were fox-like predators that likely pursued small mammal prey. Predation pressure, and perhaps some other factors, seems to have enforced a protracted period of nocturnality on Mesozoic mammal evolution known as the "nocturnal bottleneck", and we still see evidence of this adaptive shift in mammal anatomy today. Mammals, including humans, have generally poor color eyesight and are relatively susceptible to eye damage from ultraviolet light, but have contrastingly excellent senses of hearing and smell, as well as supreme tactile abilities on account of hair and whiskers. Broadly, we can view this as lessening our reliance on light and vision in favor of senses that are light-independent. We also have a unique type of fat—brown adipose tissue—that excels at rapidly warming us in cold conditions, as well as the thermal advantages of fur and high-energy metabolisms. Both are well suited to activity during cold nighttimes. Nocturnal habits are common enough in living mammals for us to assume that day-based behavior is an "advanced" trait that developed as we explored the adaptive potential of the mammal-dominated Cenozoic Era.

We should not assume that the diminutive forms and moonlit scurrying of Mesozoic mammals makes them uninteresting creatures. Our occupancy of the night was as much about exploiting nocturnal opportunities as it was about avoiding dinosaur predators. We now know that Mesozoic mammals were not just ground-based insectivores: they were also gliders, swimmers, burrowers, herbivores, and even dinosaur predators. Mammals did not spend the Mesozoic on the bench, waiting for an opportunity to diversify once dinosaurs disappeared: our earliest evolution was actually complex and inventive, and already hinting at the adaptations and habits that mammals would enhance further when opportunities arose in the Cenozoic.

Yutyrannus, *the Feathered Tyrant (Cretaceous)*

WHILE NO DOUBT REMAINS THAT MANY SMALL, BIRDLIKE THEROpod dinosaurs were feathered, much remains to be learned about the evolution of feathers among other dinosaurs. Fossils show that many dinosaurs had scaly skin over all or most of their bodies, but new discoveries have revealed feather-like structures and fibers on dinosaurs many evolutionary miles from the dinosaur–bird lineage. Presently, the fossil record is sufficient to hint at a complex picture of feather evolution in Dinosauria without being detailed enough to firmly answer when feathers and their anatomical precursors first appeared, how many times they evolved, and how they varied in response to factors like body size, habitat, and climate.

Currently, the largest dinosaur we know of with feather-like structures is the Early Cretaceous Chinese tyrannosauroid *Yutyrannus hauli.* *Yutyrannus* was 9 m long and weighed more than 1 tonne. Although far from the largest predatory dinosaur (the biggest dinosaur carnivores exceeded 14 m in length, with body masses surpassing 6 tonnes), it is much bigger than any other known feathered species and demonstrates that fibers or feathers could be retained in animals of large size. Several *Yutyrannus* specimens show long, dense filaments across their bodies, including parts of the neck, upper arm, torso, and tail. With such a broad distribution, it's likely that most of *Yutyrannus* was covered in some sort of feather-like structure. *Yutyrannus* is part of the lineage that would eventually give rise to the famous tyrannosaurid dinosaurs, including *Tyrannosaurus, Tarbosaurus,* and *Albertosaurus.* Curiously, the little skin data we have from these Late Cretaceous tyrannosaurids suggests a reduction of feather-like features: scales are known to have occurred on their faces, necks, and bellies; over their hips; and on their tails. We cannot be certain that these large, advanced tyrannosaurids were entirely devoid of filaments and fibers, but they do not seem to have been as fluffy as *Yutyrannus* or their other theropod ancestors. Does this imply that *Yutyrannus* is close to the body size limit for extensive feather-like coats, perhaps because larger dinosaurs risked overheating under a dense covering of fluffy skin? Or might another factor, such as climate, account for this difference? We are learning that many locations and time periods in the Mesozoic were not subjected to the warm greenhouse climates once thought to be ubiquitous throughout dinosaur evolution, and this further complicates our attempts to understand the adaptive significance of dinosaur feathers. The Lower Cretaceous forests that *Yutyrannus* called home had an average annual temperature of 10°C, far cooler than the Late Cretaceous floodplains and woodlands occupied by *Tyrannosaurus* and kin (annual average temperature of 18°C). Is this enough of a difference to imply that climate, as well as body size, was influencing the evolution of dinosaur skin and feathering? It's a possibility, but we need more fossils to know for sure.

Early Cretaceous tyrannosauroids such as *Yutyrannus* were just one type of predatory dinosaur among many, but by the end of the Cretaceous, tyrannosaurs were the dominant large-bodied land carnivores. All tyrannosauroids are characterized by their particularly strong skulls and elevated bite forces, but only the Late Cretaceous species have those famously short arms. Despite their length, tyrannosaurid arms were not weak: they were robustly built and powerfully muscled, and they may have been useful for gripping other animals, such as prey items or mates during copulation. Most tyrannosauroids had long legs, and the longest are found in the tyrannosaurids: despite their size, they were likely relatively swift, agile animals. The tyrannosaurids were bone-crushers, making them the only theropods with jaws and teeth powerful enough to break into large dinosaur skeletons. That they routinely injured each other, as well as their prey, is recorded in bite marks and other facial wounds preserved on their fossil skulls. This behavior is also evidenced in other large theropods, so it may have been a common means for big dinosaur predators to interact socially or to settle disputes.

Flowers and Insect Pollination (Cretaceous)

MESOZOIC FLORAS WERE MOSTLY DOMINATED BY GYMNOSPERMS, the seed-bearing plant group that includes conifers, ferns, and cycads. Vast forests and plains of these plants existed throughout the Mesozoic, supporting the evolution of some of the largest herbivorous animals of all time. To modern eyes, these landscapes would have looked overwhelmingly green, as they lacked the color associated with the flowering plants—the angiosperms—until the Cretaceous Period.

Our cultural association of flowers with romance and delicacy might lead us to imagine their evolution as equally tender and gentle, quietly blossoming color and fragrance into Mesozoic scenery. In fact, angiosperms rose aggressively and rapidly, transitioning from holding minor ecological roles in Early Cretaceous floras to being the dominant plants of terrestrial communities by the end of the Mesozoic. Their sudden abundance had significant effects on Mesozoic biospheres and climates. By virtue of their productivity and an increased capacity to transmit groundwater into the atmosphere, angiosperms enhanced the planet's capacity for rainfall, which, in turn, lead to greater erosion rates and more nutrients washing into marine ecosystems, heightening marine productivity. Today, something like 350 thousand species of angiosperms exist on Earth. Forget about living in the Age of Mammals: we live in the Age of Flowers.

We have much to learn about the early evolution of angiosperms. When they first appeared is a matter of contention, but they were probably—according to our evolutionary models and some intriguing pollen-like fossils—existent in the Triassic, if not before. Genuine flowers seem to have first appeared in the early Cretaceous, although we might note that angiosperms were not alone in experimenting with flowerlike structures at this time: several non-angiosperm lineages were evolving bright or pungent structures that may have served similar reproductive purposes. Why angiosperms had such sudden success in the Cretaceous has been a long-held mystery, but current research points to increased productivity as their chief advantage. Their enhanced capacity for growth seems to be related to a microscopic, seemingly trivial component of their anatomy: small genome size. Smaller volumes of genetic material equate to smaller cells, allowing for tighter packing of veins and pores into leaves. This is the plant equivalent of having a supercharged lung and circulatory system, and it may have given angiosperms a physiological advantage over their competitors.

Flowers may also have played a role in angiosperm success. Flowers distribute pollen—the plant equivalent of sperm—on animals that interact with them, and if animals can be coaxed into routinely interacting with the same flower species—via attractive colors, shapes, odors, or foodstuffs—plants can transform their visitors into reliable, efficient agents of fertilization. This is a much more dependable reproductive system than wind carrying pollen through the air, where fertilization depends on fortuitous gusts carrying pollen grains to receptacles on other plants. Insects acted as pollinators for plants well before the rise of angiosperms (exactly when they adopted this role is contested; there are indications that this relationship may have begun in the Paleozoic), but flowering plants ran further with this relationship than any other plant group. Fossils show that early flowers were generalist in structure and open to several types of pollinator, but by the Late Cretaceous many flower species were selective about their insect partners, requiring specialized feeding apparatuses to access their nectar. The close evolutionary kinship and codependence we see between certain plants and insects in modern times is a continuation of a very ancient phenomenon.

By the end of the Cretaceous, insect pollination was the dominant mechanism of plant reproduction, as it remains today. Indeed, for all our technological innovation, we still rely on insects to fertilize our crops. It's easy to overlook bees, moths, butterflies, and hoverflies busying themselves among our plants, or even to regard them as pests and annoyances, but the pollinating services these tiny creatures provide are absolutely essential to our way of life.

Cretoxyrhina *and* Pteranodon *(Cretaceous)*

WE TEND TO PORTRAY THE MESOZOIC ERA AS THE TIME OF GREAT sea reptiles—ichthyosaurs, plesiosaurs, mosasaurs, and so on. While there is little doubt that these animals were important to Mesozoic marine ecosystems, we should not overlook the significance of a more familiar, and yet much more ancient, group of predators that flourished alongside them: the sharks.

Sharks are one of the greatest successes of vertebrate evolution. Their fossil teeth are found in abundance all over the world through the last four hundred million years of the geological record. Fossils of their cartilage skeletons are much rarer, however, and only a few sites of exceptional preservation provide more complete insights into ancient shark anatomy. The chalk sediments deposited by the Western Interior Seaway, a shallow marine sea that bisected North America in the latter half of the Cretaceous Period, are one rock type that yields them. The shark fossils here are excellent and abundant, and they leave no doubt that sharks were an important component of an ecosystem also occupied by plesiosaurs, mosasaurs, and predatory bony fish.

Shark fossils of particular interest from the Western Interior Seaway are those evidencing their feeding on other animals. Paleontologists go to great lengths to reconstruct the diets of fossil animals, and this is aided enormously by bite marks or imbedded teeth in the bones of prey species. Western Interior Seaway shark teeth are frequently preserved in close association with the ancient carcasses of virtually all large vertebrates from this environment, and their bite marks, tooth gouges, and embedded teeth are frequently found on fossil bones. It seems that few animals in this inland sea escaped the jaws of sharks, and we know that they even turned cannibalistic on occasion: sharks eating other sharks. The 2–3-m-long shark *Squalicorax falcatus* has left particularly pervasive evidence of its foraging habits: its teeth and feeding traces are associated with so many carcasses that it was surely something of a scavenger, eating anything it could find regardless of animal type or size. A larger shark, *Cretoxyrhina mantelli* (shown here), roamed the same waters. At 6–7 m long, it was a top carnivore of the Western Interior Seaway, and fossil evidence shows that it dined on even relatively large marine reptiles.

Among the rarest quarry of *Cretoxyrhina* was *Pteranodon*, a flying reptile famous for its large, backward-pointing cranial crest; toothless jaws; and wingspan of up to 7 m. In fact, most *Pteranodon* individuals were much smaller than this, with wingspans of 3–4 m, and they had much smaller cranial crests. It is thought that these smaller morphs were females, and the males were larger and full-crested. The bones of *Pteranodon* were, like all large pterosaurs, occupied by bony air sacs, so that the bone walls were just a millimeter or so thick. This made for a lightweight and flight-adapted skeleton, but it rendered pterosaur bones highly vulnerable to damage once their owners died. Evidence of carnivorous acts on pterosaur bodies is therefore rare, but we know that pterosaurs were eaten by fish, dinosaurs, ancient relatives of crocodylians, as well as sharks. Both *Squalicorax* and *Cretoxyrhina* are known to have eaten *Pteranodon*, with our limited data set showing *Squalicorax* dining on pterosaur meat more regularly than *Cretoxyrhina*. Unfortunately, bone-imbedded teeth and bite marks rarely give clues about whether animals were being preyed upon or scavenged, but both of these shark species would vastly outweigh the largest *Pteranodon* and probably could have easily overpowered a waterborne individual. We know that *Pteranodon* ate fish, and they may have routinely entered water to catch them. Though pterosaurs were likely strong swimmers and adept at taking off from water, this behavior would place *Pteranodon* in the same habitat as these dangerous fish. Unless they were alerted to these predators and quickly became airborne, *Pteranodon* may have been vulnerable to attacks from swimming carnivores.

Mighty Zuul, Destroyer of Shins (Cretaceous)

AVOIDING THE SHARP END OF A PREDATORY SPECIES IS A PRESsure that most animals face at one time or another. Common adaptive responses include being fast enough to escape pursuit, being strong enough to fight off attackers, or simply being so awkward and difficult to subdue that the predation effort is not worth the reward. Great body size is one mechanism of avoiding attack—most predators know their weight class and won't punch far above it—but deterrents such as armor and spikes achieve the same goal without the pressure of becoming the largest creature in a given habitat. Armored animals have developed numerous times in all vertebrate groups, and they occur in a surprising variety of niches. Though armor is heavy and slows animals down, speed need not be a concern for species that eat plants, raid insect nests, or restrict their rapid locomotion to short bursts. These animals can take their time with whatever they encounter, safe in the knowledge that their defenses will frustrate or deter hungry predators.

Among the most spectacular armored animals of all time are the ankylosaurian dinosaurs. They are, along with the famous stegosaurs, part of the armored herbivorous dinosaur group known as Thyreophora. Ankylosaurs were common members of dinosaur faunas during the Cretaceous, and they developed amazing defensive structures with spines, knobs, plates, and tail clubs. The entire group is made up of wide-bodied, huge-bellied, and low-slung animals that must have trundled around at relatively low speeds, using their beaked jaws and small, coarsely serrated teeth to graze plant material. We recognize at least two categories of these dinosaurs: ankylosaurids and nodosaurids. Nodosaurids are characterized by their large shoulder spikes and narrow faces, while ankylosaurids have substantial tail clubs, heavily armored faces, and broad muzzles. It is thought that having narrow beaks enabled nodosaurids to feed selectively, while having wide beaks allowed ankylosaurids to grab mouthfuls of whatever vegetation they happened across. Enlarged nasal cavities indicate that ankylosaurs had an enhanced sense of smell, but the complex layout of their nasal apparatus suggests their noses had other functions as well, such as augmenting vocalization or controlling heat and water exchange.

The function of ankylosaurid tail clubs has drawn much attention. Like the rest of their armor, the clubs are made from bones that grow within the skin and were likely covered with thick scales and cornified skin in life. Tail clubs often vary in size and shape within ankylosaur species, implying that their main use may have been to settle contests between individuals of the same species instead of deterring attack from carnivores. Perhaps ankylosaurids primarily used their tails to batter each other in contests for mates or resources? Of course, this does not rule out additional roles in predator defense, and studies show that the largest tail clubs may have been capable of smashing large bones when swung at attackers. Indeed, the species name of the large-clubbed ankylosaurid depicted here—*Zuul crurivastator*—means "destroyer of shins."

Well-preserved ankylosaur remains easily rank among the most fantastic of all dinosaur fossils. Their densely armored backs give the impression of seeing skin rather than bones, so especially well-preserved and complete specimens resemble sleeping animals, not long-dead fossils. Excellent preservation of their bodies is facilitated by a quirk of ankylosaur fossilization, whereby their tough carcasses were capable of remaining intact and drifting far out to sea before sinking and fossilizing. Marine settings are often more conducive to fossilization than terrestrial environments, allowing these amazing animals to be gently buried in fine muds to emerge, as sleepy-looking skeletons, millions of years later.

Giant Sea Lizards and the Mesozoic Marine Revolution (Cretaceous)

AT FIRST GLANCE THE MOSASAURS, A CRETACEOUS GROUP OF aquatic reptiles, seem alien enough for us to assume that their origins lie among some long extinct, exotic lineages. In fact, details of skull anatomy show that these paddle-limbed reptiles were members of the lizard group, and were most likely close relatives of snakes and monitors. Like modern lizards, they had scaly skin, bore forked tongues, and had additional rows of teeth on the upper surface of their mouths. But unlike modern lizards, these reptiles could grow to enormous sizes, some even becoming whale-sized creatures of 15–17 m long. They occupied predatory niches in Late Cretaceous seas at a time when other marine reptiles— such as plesiosaurs and ichthyosaurs—were in decline. Gut content, as well as studies of mosasaur skull mechanics, suggests they were powerful carnivores that essentially ate whatever they wanted. Half-digested bones of birds, sharks, bony fish, and other marine reptiles have been found within their fossilized guts. Ammonites were also common mosasaur prey, with some mosasaur species being specifically adapted for crushing mollusk shells. That ammonites could escape these predators is evidenced by healed tooth marks left in their shells—wounds that have been linked incontrovertibly to mosasaur dentition. *Globidens dakotensis*, the 6-m-long mosasaur from North America shown here, apparently specialized in such prey. Unlike the sharp, pointed teeth of mosasaurs that were adapted for gripping slippery, fleshy prey, *Globidens* teeth were blunt and subspherical, and ideal for crushing shells.

Mosasaurs have long been regarded as very lizard- or crocodylian-like in appearance and swimming behavior, moving through water via strong lateral undulations of their bodies, and propelled with a frill lining the top of their tails. But recent discoveries have forced reinterpretation of both their life appearance and their swimming mechanics. At least some mosasaurs bore a well-developed tail fin, more like that of an oceangoing shark than a crocodylian, and their bodies were chunky, streamlined, and mostly stiffened, making oceanic fish or whales better functional analogues than any living reptile. These features present mosasaurs as far more aquatically adapted than historically envisaged, and we should imagine them as the lizard equivalent of ichthyosaurs or whales rather than overgrown monitor lizards paddling out to sea. Rare fossils of extremely young mosasaurs suggest that they were born at sea and did not hatch from eggs on land, another indication of how far mosasaurs took their aquatic adaptations.

We take it as a given that animals such as *Globidens* were able to smash open shellfish to access the flesh within, but the capacity to bypass the defenses of shelled invertebrates was a major Mesozoic innovation that transformed marine life. This widespread evolution of crushing, drilling, and rasping apparatuses is known as the "Mesozoic marine revolution," and it allowed certain vertebrates (fish and swimming reptiles) and invertebrates (snails and decapod crustaceans) to predate species that were previously inaccessible inside shelly homes. This forced rapid adaptation, or else quick evolutionary demise, for the affected prey species. Throughout the Mesozoic animals responded to these newly equipped predators by strengthening their shells, finding new ways to take refuge, or relocating to safer habitats. Lineages that couldn't adapt in one of these ways diminished in diversity and abundance. Brachiopods and the stalked crinoids were among those that struggled to combat shell smashers, drillers, and raspers, so they retreated into deeper, quieter habitats where predation pressures were lessened. Snails and bivalves responded more readily by thickening their shells, growing antipredator defenses such as spines, developing burrowing habits, and evolving quick escape tactics. This revolution played a major role in transforming shallow sea faunas from older Paleozoic-style ecosystems to more recognizable modern marine communities.

A Colossal Ammonite (Cretaceous)

MANY OF THE SPECIES WE'VE ENCOUNTERED THUS FAR ARE known from rare fossils, maybe even single specimens. They're the sort of creatures we dream of finding when traversing fossil outcrops, but only a tiny fraction of us are so lucky. This does not apply to the subject of our next painting, however: an ammonite. Remains of these animals are extremely common in Mesozoic marine beds, and in some localities it's easy to collect dozens or even hundreds in just a few hours. The abundance, extraordinary diversity, and rapid evolution of ammonite species makes them useful fossils for dating Mesozoic rocks. Many ammonites are characteristic to a time interval of one million years or less, so if we can confidently identify an ammonite species, we obtain a precision date for the rock it was found in.

Ammonite fossils are calcium carbonate shells coiled in a single plane, and it's this coiling that reveals their relationship to other animals. They are part of Mollusca, and specifically a type of cephalopod—the group that today contains octopuses, squid, and nautilids. Cephalopods are tentacled, free-swimming, intelligent mollusks that have an ancestry extending into the Cambrian. Shelled forms, like nautiloids and ammonoids, were common throughout the Paleozoic and the Mesozoic, but modern cephalopod diversity is dominated by shell-less forms. Only a few living species of nautilids have shells, of which *Nautilus pompilius* is the most famous.

Like nautilids, ammonite animals lived within their shells. Their bodies inhabited a chamber at the large end of the spiral, and the preceding chambers were filled with air or flooded with fluid as a means of buoyancy control. Adjusting the ratio of fluid to air in these spaces makes shelled cephalopods negatively or positively buoyant, allowing them to effortlessly ascend or descend through the water column. But though nautilids and ammonites share some basic principles of shell anatomy, ammonites generally differed from *Nautilus* in form and habits. Ammonites are actually more closely related to octopuses and squid than nautilids, and they are distinguished from living and extinct nautiloids in several ways. Modern nautilids are deep-sea creatures with eighty to ninety tentacles, leathery hoods covering the opening of their shells, simple eyes, and—unlike squid or octopuses—no ability to project clouds of ink when startled. Ammonites, in contrast, often lived in shallow waters, used part of their jaw apparatus to close their shells, and were equipped with defensive ink sacs. They may—as closer relatives of squid and octopus—have also borne fewer tentacles (squid and octopus have ten and eight tentacles, respectively) and had better eyesight than nautilids. The precise form of the ammonite animal remains unknown (a somewhat surprising fact, given the millions of ammonite shell fossils we have), but there is no reason to think they were *Nautilus* clones.

The lifestyles of ammonites remain mysterious. Their jaw apparatus was a parrotlike beak that seems suited for eating soft prey, and some ammonite fossils contain the remains of their last meals—small planktonic animals, such as crustaceans. But how they obtained their prey, and where and how they preferred to live, is not easily predicted, a conundrum made all the more complicated by the staggering diversity of ammonite shell form. In addition to the familiar coiled shape, ammonites can be intricately ornamented, uncoiled, twisted and knotted, and all sorts of variations between. It is thought that vast size differences existed among ammonite genders, with smaller, often more elaborate males being dwarfed by larger females. Size differences are also extreme between species: some shells never exceed a few centimeters across, while *Parapuzosia seppenradensis*—a species from Late Cretaceous Germany (shown here)—reached 2–3 m across the shell and probably weighed well over 1 tonne. There is likely not one answer to the question of ammonite habits, their different forms surely reflecting contrasting adaptations to varying lifestyles.

Giant Flying Reptiles (Cretaceous)

IF WE'RE EVER ABLE TO RESURRECT SPECIES FROM MILLIONS OF years ago, the pterosaurs—the first vertebrates to achieve powered flight—are surely at the top of our wish list. Using a unique membranous wing anatomy where the fourth finger—the equivalent our ring finger—supported much of the airfoil, pterosaurs flew through most of the Mesozoic after evolving, under currently poorly understood circumstances, in the Triassic Period. Their relationship to other reptiles has long been mysterious on account of their geologically oldest fossils being true, bona fide pterosaurs without obvious anatomical links to other Triassic reptiles. Decades of interrogating their fossils, however, have enabled researchers to peel back the outwardly unusual anatomy of pterosaurs and find features linking them to archosaurs, with dinosaurs likely being close relatives.

For much of their research history, pterosaurs were regarded as evolutionary failures: reptilian gargoyles that struggled to walk, could barely fly, and only served to keep the sky warm until superior fliers—birds and bats—could take over their niches. This view has been overturned in recent decades as researchers have gained greater understanding of pterosaur anatomy and biology. Their lanky proportions reflect expanded, air-filled bones that probably linked with an efficient, birdlike lung system. Their fossil footprints show that, far from having the terrestrial capability of a collapsed tent, pterosaurs strode and even ran confidently. Many species probably spent a lot of time this way, searching for food on the ground. Other lineages were adapted to chase aerial insects, to scavenge carcasses, to swim after or glean fish from the seas, or to wade through shallow waters using a variety of jaw and tooth shapes to detect and strain food. Large shoulders with expanded spaces for flight muscles indicate a capacity for powerful flapping and genuine powered flight, not simple gliding. Our perception of their diversity continues to expand too, with a glut of new discoveries—particularly in Brazil and

China—revealing unusual, fascinating new species. These same localities have yielded the first pterosaur eggs and embryos, which show that pterosaur hatchlings were so similarly proportioned to their parents that it's near certain they could fly very soon after hatching.

But perhaps the pterosaurs most sorely lost to Deep Time are the giants. Cretaceous pterosaurs routinely achieved wingspans of 3–6 m, comparable to the largest birds that have ever existed, and a number of species became even larger: behemoths with 10-m wingspans and 200–350-kg body masses. These animals belonged to a very successful, globally distributed group of toothless pterosaurs known as the azhdarchids (shown here). The fossil record of giant azhdarchids is relatively poor compared to that of their smaller cousins, but every component of their anatomy is consistent with flying habits, and flight models predict superior soaring skills that would enable them to travel thousands of kilometers with ease. The biggest pterosaurs were capable of surpassing the size of the largest birds because of their more powerful takeoff mechanism, which incorporated power from all four limbs, rather than just their legs. All animal flight begins with a leap rather than a flap, so the power available for this initial bound into the air is a major factor in determining the maximum size of a flying species. Launching quadrupedally is so efficient because it uses the largest muscles in the body, the flight muscles of the wings, to primarily power the takeoff leap, whereas bipedal takeoff—as practiced by birds—is entirely reliant on the lesser muscles of the hind limbs. Many bats use the same quadrupedal launch mechanic as pterosaurs, but as mammals they lack the expansive body air sacs and hollow bones that are probably also essential to becoming aerial giants. Pterosaurs combined their large, lightweight bodies with an efficient, powerful launch strategy to become the largest flying animals of all time. We can only imagine what it would be like to see these marvelous animals flying overhead.

Deinosuchus, *an Enormous Alligatoroid (Cretaceous)*

CROCODYLIANS ARE OFTEN PRESENTED AS LIVING FOSSILS OR even modern dinosaurs, neither of which is true. Crocodylians and dinosaurs are both archosaurs, but they have not been part of the same evolutionary line since the Triassic Period—over two hundred million years ago. The crocodylian branch of Archosauria is known as Pseudosuchia, and it has a significant role in reptile evolution. Pseudosuchia was once far more diverse than it is today, including reptile types that vied with other species on land and in water for predatory and herbivorous niches. Our modern crocodylian group was a surprisingly late addition to this roster, appearing only toward the end of the Cretaceous Period. Crocodylians would eventually dominate pseudosuchian diversity, and in the modern day they represent the last of this great reptile line. But the semiaquatic, predatory habits of living species are just one of many lifestyles practiced by pseudosuchians in their long history. Far from being living fossils, modern crocodylians are the tip of a huge evolutionary iceberg.

Among the first, and surely most spectacular, of the true crocodylians was the giant Late Cretaceous alligatoroid *Deinosuchus*. Two species of this animal are generally recognized as having lived around the coasts and estuaries of the North American Western Interior Seaway. The eastern species, *D. rugosus*, grew to 8 m long, a little larger than the longest-living crocodylian (the saltwater crocodile, *Crocodylus porosus*), while the western species—*D. riograndensis*—attained lengths of 10 m or more, making it one of the largest pseudosuchians to ever have lived. Only the Cretaceous crocodyliform *Sarcosuchus imperator* and Miocene caiman *Purussaurus brasiliensis* challenge *Deinosuchus* in size, though the lack of complete skeletal remains for any of these species precludes determination of the true record holder.

The appearance of *Deinosuchus* is often mischaracterized in art, both reflecting assumptions that *Deinosuchus* was just a scaled-up crocodile as well as an early, erroneous skull reconstruction based on fragmentary material. Often restored with a triangular skull and nothing but conical teeth, *Deinosuchus* actually had a much broader skull with sophisticated dentition: conical, piercing teeth at its jaw tips and crushing teeth further back. All its teeth were covered with thick, wrinkled enamel that reinforced them against powerful biting. A further distinction from living crocodylians are the large, bulbous, and deeply pitted scutes along its back. These were embedded in tough skin and muscle, and they served to both armor the back and strengthen it during terrestrial locomotion, a scute function shared with living crocodylians. So characteristic are *Deinosuchus* teeth and scutes that they are identifiable even when discovered in isolation, a happy fact given that they represent the overwhelming majority of this animal's fossil record. A complete picture of *Deinosuchus* proportions remains elusive thanks to its remains being damaged and scattered by storms prior to fossilization, a process that has left us with scant remains and little in the way of associated skeletons.

As an alligatoroid, *Deinosuchus* is closely related to living alligators and caiman. Like these species, it seems to have had very high biting strength. Large turtles seem to have been among its most common prey, and the fossil bones of these shelled reptiles are often riddled with *Deinosuchus* tooth marks. *Deinosuchus* evidently crushed their shells with some force, its teeth often bearing cracks and chips from powerful chewing and smashing activity. Some turtle remains show evidence of healing from their *Deinosuchus* attacks, indicating that *Deinosuchus* preyed upon live animals and was not merely scavenging dead ones. Popular culture has a penchant for the idea of *Deinosuchus* ambushing shore-based dinosaurs, and rare fossils hint that events of this nature may have happened. Indeed, one dinosaur limb bone is known that shows extensive *Deinosuchus*-induced damage, its round cross-section being mashed and chewed into a roughly square one. It was a formidable animal indeed that could use dinosaur bones as chew toys.

Triceratops (*Cretaceous*)

THE HORNED DINOSAURS, OR CERATOPSIANS, WERE AMONG THE most spectacular of all dinosaurs—no mean feat considering the number of iconic and charismatic species in the dinosaur evolutionary canon. Ceratopsians arose in the late Jurassic Period and they were represented by small bipedal dinosaurs distributed across Europe, Asia, and North America until the Late Cretaceous. As the Mesozoic drew to a close, ceratopsians developed into an astounding diversity of primarily North American, large-bodied quadrupedal forms, all of which have fantastic cranial frills, bosses, or horns. Skulls are the most common fossils of these larger species, and their ornamentation has proved very useful for distinguishing different species. By the close of the Cretaceous, horned dinosaurs were abundant herbivores in some dinosaur ecosystems, their fossils being far more common than those of other dinosaur species. Bone beds containing the remains of numerous individuals are known for several species, which implies that at least some horned dinosaurs engaged in gregarious behavior. Their apparent sociality and horned faces make it difficult not to imagine large horned dinosaurs as a dinosaurian take on cattle.

Triceratops is not only the most famous ceratopsian, but also one of the most famous extinct animals of all. A pair of brow horns discovered in 1887 was the first evidence of this genus. Partly as a result of confusion about the age of the rocks the horns were found in, these were initially thought to belong to a strange extinct bison, but the true nature of *Triceratops* as one of the last of the horned dinosaurs soon emerged. Today, it is regarded as one of the best-represented dinosaur species, with many dozens of skulls and skeletons known, ranging from small juveniles to gnarled, mature adults. Once thought to contain multiple species, only two *Triceratops* species are now recognized. *T. horridus* is shown here.

At 9 m long and with an estimated body mass exceeding 6 tonnes, *Triceratops* was among the largest of all horned dinosaurs. Its skull is characterized not only by the three large horns that give us its name ("three-horned face"), but also by a relatively short and simple frill.

Several functional explanations for ceratopsian horns and frills have been proposed, the most common being defense and sociosexual display. *Triceratops* growth series show that, compared to adults, juvenile individuals had much smaller horns and stunted frills. Given that juveniles are far more vulnerable to predation than adults, this casts doubt on a function purely related to predator defense. The development of fully realized horns and frills in adults is consistent with a sociosexual role: it's adults, after all, that tend to be most interested in competition for territory and reproductive rights. Healed wounds on *Triceratops* skulls, which correlate precisely to simulated "horn lock" combat between mature individuals, are direct evidence that *Triceratops* horns had a role in contests between individuals. Evidently their facial ornamentation was not just for show, and, of course, if *Triceratops* could battle each other with their horns, they might have used them against predatory species as well.

The life appearance of *Triceratops* is unusual compared to that of other horned dinosaurs. Some ceratopsians had a series of scales on their faces, but the faces of adult *Triceratops* seem to have been entirely covered in a sheathlike skin, possibly akin to that covering bird beaks or cattle horns. Perhaps this protected them against facial injury? Skin impressions from *Triceratops* bodies are unusual, too. The skin of ceratopsians is typically a mosaic of small scales punctuated with occasional larger oval scales. *Triceratops*, in contrast, was covered in relatively large polygonal scales, some with peculiar projections in their center. The significance of these is unclear—were they low spikes, apertures for a large bristlelike filament, or something else entirely? At least one horned dinosaur, the Early Cretaceous *Psittacosaurus*, had a row of bristles along its tail, an entirely unexpected structure in a dinosaur group with an extensive record of scaly skin. Perhaps, as better soft-tissue fossils of *Triceratops* are found, further surprises may lie in store for our appreciation of its appearance.

The Cretaceous–Paleogene Extinction

THE MOST FAMOUS EXTINCTION EVENT IN HISTORY OCCURRED sixty-six million years ago and brought the Mesozoic Era to an end. This was the mass extinction that eliminated approximately 75 percent of plant and animal species, including the nonavian dinosaurs, ammonites, pterosaurs, and the vast majority of marine reptiles. The removal of so many species meant the post-Mesozoic world was radically different from that which preceded it: fish, mammals, and birds were able to diversify dramatically as they filled niches once occupied by reptiles and other Mesozoic forms. The K/Pg ([K]retaceous/Paleogene) extinction event was not, as it's often stereotyped, just the event that wiped out the nonbird dinosaurs: it was another radical reshuffling of Earth's biosphere.

Exactly how the K/Pg event played out remains the subject of investigation. Most of us are familiar with the fact that an asteroid was involved, but this may have been only one factor among many. Several major stresses were affecting life on Earth at the end of the Cretaceous, including falling sea levels (thus reducing the amount of shallow seas, environments where life thrives) and huge fissure eruptions in the Indian subcontinent—the remnants of which we call the Deccan Traps—were releasing enough gases and dust to impact climate and perhaps limit sunlight access. The latest Cretaceous fossil record is less than exemplary, so scientists are still trying to establish how life was impacted by these stresses. It remains controversial whether some groups, such as dinosaurs, were already suffering before the very end of the Cretaceous, or if the K/Pg event struck in their prime.

The final event of the Cretaceous was the impact of an asteroid, some 10 km in diameter, into what is now the Yucatán Peninsula of Mexico. The impact created the 180-km-wide Chicxulub crater and left a layer of iridium-rich clay across the planet. Iridium is rare on Earth but common in asteroids, so it's likely that this distinctive sediment layer records fallout from an enormous impact of extraterrestrial origin.

These geological phenomena coincide perfectly with the disappearance of characteristic Mesozoic fauna and flora from the rock record to suggest that, whatever detriment volcanism and sea-level fall had already delivered to Cretaceous organisms, the Chicxulub impact was the actual curtain call for Mesozoic life.

Geological data suggests that the asteroid collision was one of the grandest, most terrifying moments in the history of life on Earth. The immediate effect was an explosion equivalent to ten billion Hiroshima-grade atomic bombs. Organisms living within one thousand kilometers that were not killed by the initial blast would have encountered massive energy output that triggered wildfires, earthquakes, the collapse of the Yucatán continental shelf, and tidal waves hundreds of meters tall. Organisms even five thousand kilometers away would have been showered with dust and particles ejected from the impact site, burying landscapes in debris layers up to 10 cm thick. Around the world, particles hot enough to start fires or kill exposed organisms rained down after having been being blasted into space, and tsunamis ravaged coastlines. As these violent events subsided, the long-term effects of the impact began, enhancing the stresses already catalyzed by the Deccan Traps eruptions. The impact had vaporized rocks rich in carbon and sulfur, turning rainfall acidic and filling the atmosphere with sunlight-scattering aerosols. These aerosols reflected sunlight away from Earth's surface and began to cool the planet, an effect worsened by sun-blocking dust blown into the atmosphere during the impact. With Earth's surface receiving only weak sunlight, the entire planet cooled by 10°C. Earth remained dim and cold for decades, weakening ecosystems that were adapted for receiving plentiful solar energy. Only plants adapted to low light and detritus-based food chains thrived in this interval, while forests, complex marine ecosystems, and large animals disappeared from the planet. Recovery to pre-extinction diversity took, in some ecosystems, up to three million years.

House Mice and Other Modern Evolutionary Winners (Holocene)

WE HAVE ALREADY SEEN, COURTESY OF *MORGANUCODON* AND some Mesozoic mammals, that a fairly generalized small-bodied form was a proven success for mammal-line animals early in their evolution. Rodents—the most speciose modern mammal group (2,277 living species, comprising about 40 percent of all mammal diversity)—show that this body plan remains useful today. Rodents are extremely adaptable and have colonized most continents in great numbers; Antarctica is the single holdout against their spread. Though most rodents are small—rats, mice, hamsters, squirrels, guinea pigs, and so on—some living examples are the size of large dogs (beavers, porcupines, and capybaras). In the past, rodents experimented with being even bigger, including South American forms as large as rhinoceroses, and beavers the size of humans. They are undeniably one of the greatest successes of mammal evolution, true champions of the post-Mesozoic "Age of Mammals."

Rodents are specialist chiselers with two pairs of continuously growing incisor teeth and powerful jaws. This enables them to gnaw at foodstuffs that would be too tough for other animals to eat, as well as to manipulate their environment: chewing into structures to create dens or nests, or felling and gathering robust building material for their own architecture. Their characteristic jaws and teeth make their fossils highly identifiable, and we can trace the evolution of the four major rodent groups into rocks of the Paleocene Epoch. Some disagreement exists, however, over whether these remains represent the first rodents. While fossils suggest that rodents arose in the aftermath of the Cretaceous mass extinction, genetic data predicts a slightly older origin, just before the end of the Cretaceous. In either eventuality, rodents were well equipped to capitalize on the potential presented in early Cenozoic ecosystems, and today they are among the most abundant mammals on the planet. As fast breeders adapted to giving birth to large numbers of offspring, rodent populations increase rapidly and spread quickly. The oldest fossil rodents indicate an origin in China and Mongolia, and within a few million years they had spread to Europe and North America. By the middle Eocene they had rafted over the Atlantic Ocean from Africa to South America (then isolated from other continents), and they finally reached Australasia within the last few million years.

Though many rodent species are endangered today, a handful of species are so well adapted to living alongside humans that they are considered pests: species that parasitize our settlements by stealing food, infesting our buildings, and causing health concerns. But this does not reflect a nefarious or intentioned assault on humanity: species such as house mice (*Mus musculus*, right) are simply exploiting our way of life in the same way they would approach any evolutionary scenario, making the most of circumstances because they can. Indeed, this is the reality of all "pests": they are not species that we happen to share environments with, but those that directly benefit from the way we live. By shaping the world with extensive agriculture and urban settings, we have created conditions conducive to the likes of rodents, pigeons, and weeds, and they propagate because of us, not in spite of us. Our pests have followed us as we've colonized the globe, stowing away on ships or following the development of farms between towns and cities. Once restricted to regions of Asia, house mice and brown rats (*Rattus norvegicus*) now live worldwide, making them a major nuisance to humans as well as invasive species that endanger indigenous wildlife across the globe. Invasive rodents can spell disaster for foreign ecosystems: their breeding potential and adaptable feeding habits often overwhelm native small mammals (including other rodents) and spell doom for species such as ground-nesting birds, which are ill-equipped to deal with them. We may regard pests and their ecological impact as operating independently of humanity, but they are really an extension of our own success. They stand on our shoulders to gain an evolutionary boost, and capitalize on the chances we have provided for them.

Gastornis, *a Giant Land Waterfowl (Eocene)*

THOUGH MANY DINOSAURS BECAME EXTINCT AT THE END OF THE Cretaceous Period, one lineage continued to flourish throughout the Cenozoic: the birds. Today, they are some of the most speciose and charismatic of all animals, a group of feathered theropods that lives on every continent and in most major habitats. The Age of Dinosaurs did not end with the K/Pg extinction; it was simply retooled into an all-avian spin-off.

The roots of modern bird diversity extend into the Late Cretaceous, with many survivors of the K/Pg event representing early members of present-day bird groups. Birds underwent an explosion of diversity early in the Paleogene Period. At this time, most birds were adapted to life in open settings, likely reflecting a scarcity of extensive forest environments in the aftermath of the K/Pg extinction. Many Cenozoic avians would have looked unusual to us, taking on dramatically different anatomies, body sizes, and lifestyles compared to their closest living relatives. This includes the famous Paleogene bird *Gastornis*, a human-sized flightless bird that lived across Asia, Europe, and North America in the guise of several species. Skeletons of this heavyset bird look remarkable to modern eyes, being comparable in size to ratites (the group that includes ostriches and emus) but having robust, bulky features that are difficult to place among living species. The New Zealand flightless rails known as takahē (*Porphyrio hochstetteri*) are perhaps our best, though still crude, anatomical analogues. *Gastornis* has no close relatives among the ratites or rails, however, and is probably an early offshoot of the waterfowl branch—the Anseriformes—making swans, geese, and ducks its closest living relatives. Early reconstructions of *Gastornis* assumed a ratite-like appearance with long shaggy feathers, but their affinities to waterfowl imply a tidier appearance with neat vaned feathers. The discovery of a giant feather from a *Gastornis* site in the United States adds credence to this interpretation.

The lifestyle of *Gastornis* has been the focus of several studies. It was clearly better suited to walking than running, on account of its stout limbs and hooflike claws, and its proportionally large head implies a strong jaw for powerful biting. Computer modeling predicts a high bite strength and excellent stress distribution across its skull during feeding, but opinion has been split over how to interpret this. Was *Gastornis* a powerful carnivore, killing prey with forceful bites and maybe cracking open their bones like an avian hyena, or was it an herbivore adapted to eating tough plant material? Further studies, which include analysis of *Gastornis* jaw musculature, jaw shape, running speed, claw morphology, and bone chemistry, point to a herbivorous lifestyle being more likely. Rather than seeing *Gastornis* as the devourer of diminutive early horses, as it's portrayed in many paleoartworks, we should envisage it using its powerful beak to eat tough vegetation and nuts. In this respect, the habits of *Gastornis* were similar to those of mihirungs (dromornithids), a closely related extinct group of large ground birds that lived in Australia from the Oligocene to Pliocene, but different from the predatory phorusrhacids, which were formidable, carnivorous, and often giant terrestrial birds of North and South America.

Fossil beds composed of large, broken fossil eggshells are known from southern France that might, based on their size and geological age, represent *Gastornis* nesting sites. The abundance of these shells implies colonial nesting behavior localized to the same region from generation to generation, creating a remarkable mental image of possible *Gastornis* reproductive behavior. Growth rings in *Gastornis* skeletons suggest that they grew more slowly than most modern birds, taking several years to attain their full height of 1.5–2 m. We can assume, based on the reproductive strategies of living fowl, that *Gastornis* chicks would be precocial, likely able to follow their parents and feed themselves.

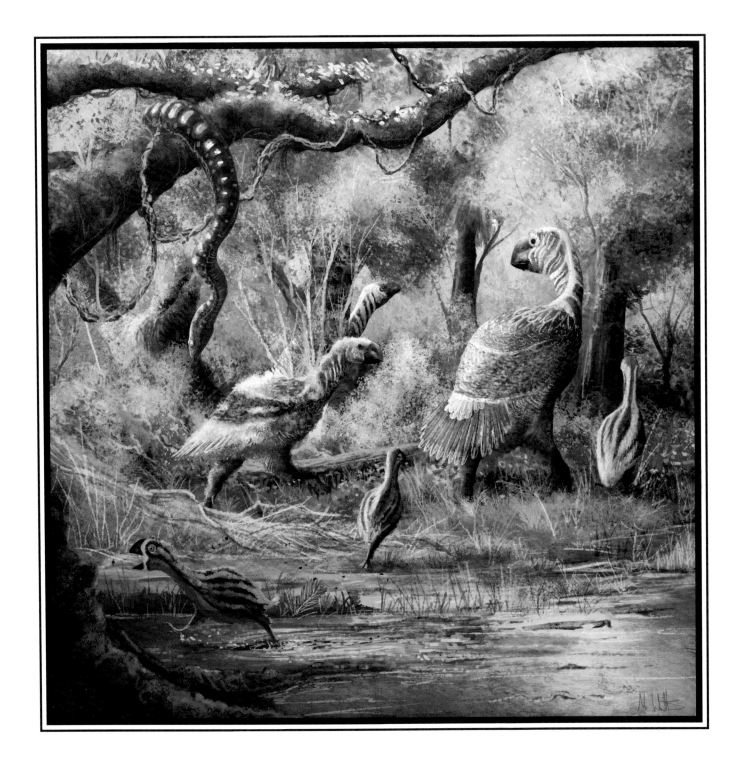

Onychonycteris, *an Early Bat (Eocene)*

THE RADIATION OF BIRDS WAS NOT THE ONLY MAJOR EVOLUTIONary event occurring among flying animals in the early Paleogene. Bats, the only known mammalian foray into powered flight, also appeared at this time. The bat group, Chiroptera, contains approximately twelve hundred species and represents about 20 percent of modern mammal diversity—only rodents are a more speciose lineage. Their ability to fly has allowed bats to spread to almost every landmass on the planet, with just the Arctic, the Antarctic, and a handful of remote islands free from their presence. The largest bats are sizeable animals with wingspans of 1.7 m, while the smallest vie for the title of most diminutive living mammal, having wingspans of just 15 cm, and body masses under 3 g. Though superficially rodent-like in appearance, bats and rodents are not closely related. Bats actually stem from the same branch of mammal evolution that houses the carnivorans and hoofed mammals.

Fossils recording the early evolution of bats are extremely rare. As with pterosaurs, our fossils representing the earliest stages of bat evolution are already entirely batlike, including a full suite of flight adaptations. Species representing previous stages of their evolution, which perhaps included flightless climbers as well as gliding forms, remain elusive. We can at least be thankful that several of our oldest bat fossils, from rocks deposited fifty-five million years ago, are well preserved, complete skeletons. The bat fossil record is otherwise largely formed of isolated teeth and jawbones, so these excellent Eocene fossils give us important insights into how bat anatomy and flight abilities have developed since the beginning of their evolutionary history.

All bats have membranous wings supported by long fingers. Their finger bones are slightly pliable on account of having reduced mineral content, permitting their wings to adopt especially aerodynamic shapes throughout the flap cycle. The wing membranes are more than just stretched skin: they contain sheets of muscle that allow the bat to control membrane stiffness in each flight stroke. These aspects mean that bats are not merely "rats with wings" but sophisticated, highly evolved aeronauts with more control over wing shape than birds and (probably) pterosaurs. They are supreme aerial acrobats, a fact exploited extensively by microbats, a (possibly artificial) group of flying insectivores numbering over one thousand species. They catch flying insects either in their jaws or with a membrane between their legs, using echolocation to find their way through dimly lit settings. Note that, while not all bat species can echolocate, none are blind.

Onychonycteris finneyi, a small (wingspan of 220 mm) fossil bat from Eocene rocks of Wyoming, United States, is among the oldest known bats and the earliest "grade" of bat evolution yet represented in the fossil record. Though undeniably bat-like in form, it retains a number of features from its ancestors that were lost in later bats, including claws on all its fingers (living bats have just one or two wing claws) and relatively long legs. Its wings have proportions typical of gliding mammals, though the presence of augmented shoulder and forelimb anatomy shows it was capable of true flight. Similarly proportioned living bats alternate between flapping flight and gliding flight, and it is possible that *Onychonycteris* did the same. Its jaws and teeth suit those of an insectivore, but its ear anatomy is unusual compared to both other fossil bats and modern species: this may indicate an inability to echolocate. We probably need superior fossils of *Onychonycteris* to be confident of this, however, meaning a long-standing question about bat evolution—flight first, or echolocation first? —remains unanswered. The long limbs and short wings of *Onychonycteris* may have allowed it to seek food on the ground or when climbing, as well as in the air. Several living bat species are similarly adapted today, their wings placing no restriction on their ability to walk or run. We do not know if *Onychonycteris* was nocturnal like most living bats, but adaptations in its feet imply an ability to hang upside-down in classic chiropteran fashion.

Arsinoitherium *and the Evolution of Giant Mammals (Eocene)*

MOST OF THE ANIMALS THAT SURVIVED THE K/PG EXTINCTION were small-bodied species that could find adequate food in resource-stripped ecosystems. As the biosphere rebuilt itself in the early Cenozoic, larger animals returned in the form of big-bodied mammalian herbivores. They initially included creatures like the globally distributed Pantodonta, which evolved in the Paleocene and developed a diverse range of omnivorous and herbivorous forms, including heavyset, 500 kg species with skulls over half a meter long. The robust, short tails in some pantodonts may have allowed them to adopt a tripodal rearing pose for browsing taller vegetation, using their tails as a prop to support their weight. Most had hoofed limbs adapted for weight bearing, although some possessed clawed digits of unknown purpose. Like some modern deer, several pantodont species bore large fangs that may—based on their lack of wear—have been used more for sociosexual display and combat than for foraging. Though appearing early in the Age of Mammals and lacking some anatomical innovations of later mammalian herbivores, the pantodonts held their own against competition from large herbivorous birds and mammals until the late Eocene or earliest Oligocene.

Pantodonts were not the only lineage experimenting with large body size in the early Cenozoic. The Embrithopoda, a group traditionally (though not exclusively) thought to be related to sirenians (dugongs and manatees) and elephants, were also Paleogene giants. The most famous and completely known member of this group is *Arsinoitherium*, a genus that occurred across northern Africa during the late Eocene and early Oligocene. It represents the last and largest of the embrithopods, with *A. zitteli* (shown here) reaching a rhinoceros-like shoulder height of 1.75–2 m, a body length of 3 m or more, and a body mass of at least one tonne. They inhabited swamps and mangrove-like settings, though their limb anatomy and bone chemistry suggest that they were land-based animals that ate terrestrial plants. This finding is surprising in light of their relatively poorly developed hips and shoulders for animals of their size, as well as their short limbs. These features have been historically regarded as indicating semiaquatic habits, but modern data suggests *Arsinoitherium* was more rhino-like than hippo-like in ecology. *Arsinoitherium* dentition and jaws suggest a powerful chewing mechanism adapted for shearing and grinding of bulky, malleable plant matter, such as large fruits. Fossils of such food items exist with abundance in *Arsinoitherium* fossil sites, perhaps representing a common component of their diet.

The most striking feature of *Arsinoitherium* are the twinned sets of horns atop its head: two large horns at the front, and two smaller ones at the back. These structures have hollow interiors and thin exterior bone walls, and their surface textures indicate that they were covered with a tough horn sheath. This approach to horn construction is very similar to that of modern bovids (goats, cows, and antelope) and is also seen time and time again in various mammal and reptile lineages throughout Deep Time. It blends a lightweight and bending resistant core (the horn skeleton) with a covering that is excellent at distributing impact forces evenly across its surface (the horn sheath). The result is a lightweight, damage-resistant structure supremely adapted for display and antagonistic behavior. With such horns, *Arsinoitherium* need not be shy about following up intimidating displays with physical aggression, perhaps locking horns with rivals or using the horns to drive away predators. Fossils show that the horns of juvenile *Arsinoitherium* were much smaller than those of adults, so perhaps their greatest role was settling disputes between older *Arsinoitherium* individuals for food, territory and reproductive resources.

Georgiacetus, *the Last Land Whale (Eocene)*

THE EVOLUTION OF WHALES WAS, UNTIL COMPARATIVELY RE-cently, not well understood. All fossil cetaceans known before the 1980s were fully marine animals entirely divorced from terrestrial habitats, and thus species largely unhelpful in answering questions about how and when whales transitioned from land to sea, or which stock of land mammals they were descended from. Light was finally shed on this evolutionary transition when fossils representing the earliest members of the whale lineage, which had strong walking limbs and lived in fresh-water habits, were discovered in India and Pakistan. In having not yet deviated markedly from the anatomy of their terrestrial ancestors, these Eocene species confirmed a long-standing hypothesis that whales had ancestry among the Artiodactyla (even-toed hoofed mammals), a finding also borne out by analysis of whale DNA. It now seems likely that, among living animals, hippos are the closest relatives of whales. Continued discoveries of Eocene whales have provided a nearly continuous sequence of whale evolution from land to sea, transforming this poorly understood portion of mammal history into one of the best-documented mammalian evolutionary transitions on record.

Cetacean evolution was rapid and was strongly directed toward the development of marine forms. It took just ten million years to evolve fully whalelike, wholly aquatic animals from the oldest known semiaquatic "proto-whales." Cetaceans began their evolutionary life looking somewhat like heavyset dogs, and they may have run or punted through water rather than swimming, their unusually heavy bones helping them to remain submerged as they pursued aquatic prey. They eventually became more crocodile-like in overall form, becoming true swimmers with long jaws and piercing teeth. But their hind limbs remained crucial to their locomotion, their tails seemingly lacking flukes and their expanded, probably webbed feet being their main propulsors. The retention of large hips with strong connections to their spines allowed their legs to support their weight on land and permitted efficient walking and running.

From this early form, cetaceans rapidly enhanced their aquatic adaptations. They developed longer jaws for seizing prey, reconfigured their teeth for a purely carnivorous seafood diet, and modified their bodies and limbs for more efficient swimming. It was among the protocetids, a group (or maybe "grade") of Eocene proto-whales found across the Northern Hemisphere, that cetaceans reached the end of their evolution on land. Fossils of pregnant protocetids show that some members of this group still gave birth out of water, their calves emerging headfirst rather than, as with living whales, tailfirst (this being an adaptation to avoid drowning during birth). But the pelvises of protocetids were joined to their spines only loosely, the connection limited to just a few vertebrae in most species, or even none in *Georgiacetus vogtlensis* (shown here). This would render their hind limbs less able to support their weight on land, as would the increasing length of their spinal columns. These issues would be compounded further by their growing size: *Georgiacetus* is estimated to been about 6 m long and might have weighed several tonnes.

We might imagine these last of the terrestrial whales as moving like large seals, undulating their bodies and using limited purchase from their limbs to clamber about beaches and exposed rocks. Protocetids retained large feet to power their swimming, but rather than paddling, they used their feet in concert with their increasingly powerful, broad tails. Animals like *Georgiacetus* were among the last whales to have possessed well-developed hind limbs and, once the need to visit land to give birth was lost, whales replaced their hind-limb flippers with a strong tail fluke. Flukes are entirely soft-tissue structures and thus do not fossilize readily, but they are detectable by the characteristic shape of their supporting vertebrae. We have yet to find such vertebrae in any protocetid, which suggests that they likely lacked large tail flukes. These structures quickly appeared in subsequent lineages after cetaceans committed entirely to marine life, however, just one of several changes that took place to fully adapt their bodies to life at sea.

Paraceratherium, *a Giant Rhinocerotoid (Oligocene)*

LIVING RHINOCEROSES ARE FAIRLY PREHISTORIC-LOOKING beasts thanks to their horns; their peculiarly shaped faces; and their thick, armored skin. Of course, they are as "modern" as any other animal lineage and not in any way archaic. To the contrary, they are quite anatomically distinct from many of their fossil relatives, and several of their characteristic features—like their horns and stout bodies—are relatively recent evolutionary innovations.

The full diversity of the rhinoceros lineage is vast, with modern rhinoceroses being the proverbial tip of their evolutionary iceberg. Rhinocerotoidea split from the perissodactyls (the odd-toed hoofed mammals, a clade that also includes tapirs and horses) in the Eocene Epoch and evolved into a myriad of forms that lived across Africa, North America, and Eurasia. Some were small, fast runners; others were similar to modern horses in size and build; some were rotund, semiaquatic creatures; and another group became the heavyset herbivores that cling to survival in Asia and Africa today. It is a sad fact that poaching will likely cause the final demise of the rhinocerotoid lineage, and quite possibly within our lifetimes. Rhinoceros are killed for their horns: structures composed of the same worthless keratin proteins as our hair, skin, and fingernails, but regarded as cancer cures or luxury commodities in parts of East Asia. At time of writing, a large rhinoceros horn is worth a quarter million US dollars in black market trade, a sum great enough to make rhinoceros horns of any kind a target for criminals. Museum taxidermies must now sport prosthetic horns; captive zoo rhinos are slaughtered in covert, nocturnal raids; and wild rhinoceroses are constantly guarded to deter poachers. It's among the wild rhinoceros that this situation becomes most desperate, as park rangers and poachers regularly exchange gunfire, and are even killed, in their quest to obtain or protect rhinoceros keratin.

Among the most impressive fossil rhinocerotoids were the indricotherines. These enormous animals roamed central Asia during the Oligocene and achieved estimated body masses of 15–20 tonnes, sizes that were outdone only by sauropod dinosaurs and, maybe, the largest mammoths. This makes indricotheres strong candidates for the largest land mammals of all time. Several different species of these giants existed, but how many, and how they are related to one another, is controversial owing to their scrappy fossil record. The most famous and largest is *Paraceratherium transouralicum*, shown here. The exact proportions of the largest species remain unclear because substantial fossil skeletons remain elusive. The history of indricothere research contains numerous different takes on their skeletal form, with some researchers suggesting they looked like giant versions of living rhinoceros, and others producing creatures that look like robust giraffes.

The reality may be somewhere in between. Indricotherines are related to (or may be part of) the rhinocerotoid Hyracodontidae, and they shared their gracility, short torsos, and long limbs. They would thus have been svelter than modern rhinos, and were probably relatively sprightly for their size. Indricotherines also had a reasonably long neck, though exactly how long remains to be determined: modern reconstructions still differ in this regard. Their skulls are well known and, though possessing powerful rhino-like jaws and massive teeth, they lack features we associate with the presence of rhinoceros horns. A further contrast with modern rhinoceros stems from indricotherine skull anatomy indicating a tapir-style proboscis at the end of their snouts. They likely used them to browse from trees, scooping vegetation into their mouths or stripping leaves from twigs. This may seem like a bold claim, given our lack of any fossilized indricothere soft tissues, but trunks and proboscises require significant reorganization of skull anatomy to house the demands of their musculature and nervous tissues, and we can identify these adaptations in fossil animals with well-preserved skulls. Altogether, indricotheres may have looked more like gigantic, heavyset horses than rhinoceros or giraffes, although we should await more definitive assessments of their anatomy and proportions before committing fully to this reconstruction.

Bluefin Tuna and the Dominance of Teleosteans (Holocene)

OF ALL MODERN FISHES, 96 PERCENT—SOME TWENTY-EIGHT thousand species—belong to one group: the teleosteans. This remarkably adaptable lineage inhabits more aquatic environments than any other animal group, including a range of marine and freshwater settings as well as inhospitable habitats such as cold arctic waters, pitch-dark caves, the deep sea, lofty mountain streams, and hypersaline lakes. The secret to this success lies in their highly adaptable bodies, which have attained a tremendous range of shapes, sizes, and feeding apparatuses to suit numerous lifestyles and environments. Teleost jaws differ from those of other fish in rapidly protruding forward during feeding, extending the reach of their mouths and sucking food inward by creating a pressure gradient in their oral cavities. Additional characteristic features are found in their tail skeletons, which are more reinforced than those of other fish and thus are superior at generating thrust for swimming. Teleosteans have applied these anatomies to almost every conceivable vertebrate lifestyle, including predation, herbivory, filter-feeding, and parasitism.

Teleosteans have their origins in the Triassic and diversified steadily throughout the Mesozoic. Most of their major body plans had evolved by the end of the Cretaceous and they became the dominant fish group in the Cenozoic, radiating at a startling rate into the thousands of species known today. Teleosteans belong to the Actinopterygii, the ray-finned fish, a large clade that has lightweight rods of bone supporting their fins instead of robust, limb-like skeletons, as were present in the early tetrapods and coelacanths we encountered earlier. Teleost skeletons are also lightweight in other respects, their bodies being composed of thin bony scaffolds instead of heavy, robust bones. This makes them relatively light and flexible compared to other swimming vertebrates, and thus generally faster and more maneuverable.

Some of the most interesting living teleosteans are also the most familiar: tuna. We mostly know these animals from their canned flesh, steaks, or sushi, but appreciating tuna only from our dinner plates does a disservice to their amazing anatomy and optimization to the niche of a fast, large-bodied predator. Species such as the northern bluefin tuna (*Thunnus thynnus*), shown here, can grow over 3.5 m long and nearly 1 tonne in mass. They are effectively fishy torpedoes, equipped with a streamlined, muscular body and a narrow, thrust-optimized tail fin. Their sickle-shaped body fins can be extended for steering and stabilization or pressed into notches on their bodies to reduce drag. Unlike most fish, tuna are warm-blooded and possess an enhanced ability to circulate oxygen and absorb it into their tissues. They are among the fastest animals in the sea and are dangerous predators to a variety of small fish, squid, and other invertebrates.

It is not only the remarkable physiology and anatomy of tuna that makes them famous: they also epitomize current, critical issues facing life in our seas. Demand for the flesh of the three bluefin tuna species has seen them overfished to the extent that they are all endangered, some critically so: the meat of some species is literally worth more than its weight in gold. Aquaculture—offshore vats full of captive fish—seems like an ideal countermeasure to their overfishing, but it is difficult to keep the pesticides, antibiotics, and other agents used in these farms from spreading to, and damaging, neighboring local marine habitats. Moreover, although farming predatory fish reduces pressure on wild populations of the livestock species, other fish species must be caught to supply the farm with feed. While this moves the risk of overfishing from one species to another, it does not remove it entirely. Oceanic food chains are long and complex, and there is real risk that poor fishing practices will, in the long-term, prove disastrous for many species, including ourselves. The vastness and productivity of our oceans places them as a potential solution to many of humanity's food crises, but far stricter management is needed if we wish to enjoy healthy, fishable marine environments beyond the immediate future, and if we want to preserve the remarkable animals that have inhabited these settings for many millions of years.

Daeodon, *a Formidable Piglike Creature (Miocene)*

DAEODON IS THE SORT OF ANIMAL THAT PREHISTORY IS FAMOUS for: a big, intimidating species that looks like it could go seven rounds with most modern species and emerge victorious. It is part of the artiodactyl group known as Entelodontidae, a clade of omnivorous mammals that lived in North America, Europe, and Asia from the mid-Eocene to the early Miocene. They bear more than a passing resemblance to pigs with their huge, gnarly skulls, large tusks, stout bodies, and hoofed feet. This similarity often sees them restored as boar- or warthog-like, and it has earned them several pig-derived nicknames among paleontological enthusiasts ("hell pigs," "terminator pigs"). But despite appearances, they are not closely related to pigs at all. Rather, they are actually part of the terrifically named group Whippomorpha: the branch of hoofed-mammal evolution that gave us whales and hippos. Unlike either of these groups, entelodonts did not live in aquatic habitats, but frequented woodlands and plains.

Two species of the North American entelodont genus *Daeodon* are known. *Daeodon shoshonensis*, shown here, is the largest, and last, of their lineage. It was a huge animal that measured 1.8 m tall at the top of the shoulders, and it typifies entelodont anatomy in having a huge skull (some 30% of the body length); a short neck; massive shoulders that helped support the weight of the head; a deep torso; and surprisingly long, slender legs. Entelodont skulls are remarkable structures with long jaws, cavernous spaces for jaw muscles, forward-facing eyes, and a number of ornamental flanges and bosses. Versatile feeding habits are indicated by their array of tooth types: massive, peg-like incisors at the front of the jaw; long, sharp tusks behind them; triangular shearing teeth behind these; and, broad, cusped teeth in the cheek region. This dentition afforded entelodonts the capacity to grip, cut, and crush their food, which was most likely plant matter (roots, tubers, fruit, and fibrous matter such as leaves and branches), as well as meat—both scavenged and predated.

Entelodonts could open their mouths exceptionally wide, permitting manipulation and strong biting action on large food items, even at the back of the toothrow. The teeth of old entelodonts are often exceptionally worn and chipped, suggesting frequent biting of hard foodstuffs. The level of wear is so extreme that it compares well to the teeth of old dogs and hyenas—animals that routinely gnaw into bone. That entelodonts were at least part-time carnivores is confirmed by their bite marks on fossil mammal bones as well as a remarkable fossil cache of seven Oligocene camel (*Poebrotherium*) individuals, all riddled with bite marks from the large entelodont *Archaeotherium*. Almost six hundred bones from several partly articulated camel skeletons were found in this association, and details of their arrangement implies that these carcasses were deliberately stored together, not randomly accumulated by water currents or other environmental phenomena. The camel skeletons are mostly left with their forelimbs and rib cages intact, while the pelvic regions and hind limbs are gone. Perhaps, like many modern predators, entelodonts prioritized eating the muscular, fleshy haunches over other parts of carcasses. A lack of feeding traces from other carnivores implies that the entelodonts killed the camels before storing them.

Bite marks on entelodont skulls are evidence that they grappled each other's faces with their mouths, and this offers another potential explanation for their wide gapes. Such behavior might relate also to the function of their cranial bosses and flanges. Perhaps an impressive set of ornaments intimidated rivals and discouraged fighting but, if conflict was unavoidable, they may have also deflected bites away from vulnerable areas. Entelodont limb proportions match those of running animals, and they were probably surprisingly fast despite their large size, massive heads, and deep chests. All evidence points to entelodonts being awesome animals, though probably ones we would be wise to avoid meeting in person.

Deinotherium, *the Chin-Tusker (Miocene)*

THE MAMMAL LINEAGE THAT GAVE RISE TO ELEPHANTS HAS A long and relatively well-understood evolutionary history. Some parts of elephant anatomies superficially resemble those of hippopotamuses and rhinoceroses, but they are not closely related; their similarities instead reflect common adaptive responses to the challenges of supporting many tonnes of body mass on land. The true closest living relatives of elephants are much less obvious: the rodent-like hyraxes and the aquatic sirenians (dugongs, manatees, and kin). Together, these groups form the clade Afrotheria. As was discussed when we encountered the giant rhinoceratoid *Paraceratherium*, characteristics of animal skulls betray the presence of trunks, and we can predict that virtually all elephant-line afrotherians had a trunk or a proboscis of some kind. The elephant lineage is thus aptly named Proboscidea.

Many fossil proboscideans have vaguely elephant-like proportions, but this was not always so. When proboscideans first split from other Afrotheres, sixty million years ago, members of the elephant line were squat creatures resembling hippopotamus-like pigs, and they were likely semiaquatic in behavior. They had highly mobile lips or short proboscises and small, forward-projecting tusks formed from oversize incisors—the same teeth that later proboscideans would grow to enormous proportions. These early proboscideans would eventually abandon aquatic habitats for a terrestrial existence, growing longer legs, great body size, longer trunks to reach the ground from their stately heights, and a variety of tusk shapes for differing adaptive purposes. In this guise, proboscideans colonized much of the planet, with only Antarctica and Oceania remaining beyond their reach.

Deinotherium is one of the first very large land proboscideans. A true giant even among the elephants and their relatives, the estimated shoulder height of some individuals approached 4 m and their masses were likely in the 10-tonne range. At least three *Deinotherium* species existed across Africa, Asia, and Europe from the Miocene to the Pleistocene.

They differ only slightly from the oldest species, *D. giganteum* (shown opposite), suggesting that the *Deinotherium* body plan was a versatile one suited to changing habitats and climates.

The face of *Deinotherium* is remarkable for a number of reasons. It evidently had a trunk of some kind, though the regions for muscle attachment are broader and longer than those of living elephants, which have tall and narrow trunk attachment sites. Some aspects of the skull imply a shorter, maybe tapir-like proboscis instead of an elephant-like trunk, although some researchers have questioned how *Deinotherium* would drink with such an organ (this, of course, assumes it *did* drink: some mammals are capable of taking all their water from their food). Adding to this mystery are two peculiar chin tusks, structures that projected backward and downward and were evidently, because of their extensive wear, used for some practical purpose. This configuration might seem unusual because modern elephants have tusks only on their upper jaws, but proboscidean history shows a great variety of tusk configurations and many species bore them on both upper and lower jaws. Evidence of abrasion between the tusks of *Deinotherium* implies that food was dragged between them, so perhaps *Deinotherium* stripped bark or leaves as a way to better prepare their food for eating, as is done by modern elephants. In all likelihood this was not the only function of their tusks: pulling over trees, fighting, and intimidating enemies are just a few other possible uses.

Though *Deinotherium* resembles an elephant, it is erroneous to think of this animal simply as an elephant with an unusual face. It has proportionally long limbs; a shorter torso; and a somewhat longer, more flexible neck. It was probably elephant-like in many ways, but its distinctive anatomies have implications for locomotion mechanics, digestive capability, and foraging techniques. These anatomical differences impact ecology and lifestyle, and they may explain how *Deinotherium* was able to live alongside other, more typically elephantine proboscidean species.

The Killer Sperm Whale Livyatan (Miocene)

WE PREVIOUSLY MET MEMBERS OF THE WHALE LINEAGE IN THEIR final stages of becoming fully marine animals. By the end of the Eocene, whales had not only made this transition but had also split into the two major groups of cetaceans we recognize today: the toothed whales (odontocetes) and the baleen whales (mysticetes). They shared the water with another type of whale, the basilosaurids, a grade of early whales that includes the famous genera *Basilosaurus* and *Dorudon*. But whereas basilosaurids perished before the end of the Eocene, odontocetes and mysticetes survived to become major predators in Earth's seas and oceans, a role they still occupy today. Some mysticetes, such as the blue and fin whales, are the largest organisms to ever have lived. They build their immense bodies from plankton, small fish, and squid that they harvest by lunge-feeding on a huge scale. While swimming at their prey, they engulf tonnes of seawater and animals in a single mouthful before their powerful throat muscles force the water out through filters of baleen—a stiff, bristlelike proteinaceous structure lining the upper jaws. Anything left in their car-sized mouths is trapped and swallowed. These amazing superpredators can eat entire schools of fish in one action.

Early marine whales did not use capture-and-filter approaches to obtain food, however. Their feeding mechanics were more similar to those of modern toothed whales: the group that includes dolphins, porpoises, and sperm whales. These animals apprehend their prey by using sharp teeth or, in some species, by suction feeding, drawing prey into their mouths by means of strong pressure gradients. Today, most odontocetes target prey species that are easily subdued, such as relatively small fish and squid. The largest of the group, the sperm whale, is famed for hunting giant and colossal squid at great depths, but these squid—which weigh hundreds of kilograms—are still much smaller than their whale predators, which routinely weigh 10–40 tonnes. Only orca (killer whales) habitually pursue large prey, and they have a reputation for being tenacious, crafty hunters. Killer whale pods will exhaust and harass other whales through long chases before eating their tongues or their calves, and they will use a variety of tactics to disable and kill seals.

Modern orcas are an echo of a time when whales generally had an eye for larger prey. During the Miocene, several species of raptorial sperm whales roamed the globe, using their massive skulls and huge teeth to prey on other marine mammals. Among the largest was *Livyatan melvillei*, a 14–17-m-long Peruvian species comparable in size to the living sperm whale. Unlike the sperm whale, however, *Livyatan* and its relatives were equipped with huge, interlocking, tusklike teeth in both jaws, those at the front being pointed for gripping prey, and those at the rear adapted for cutting and shearing. The skulls of all sperm whales have large basins that house a fatty organ known as the junk (other odontocetes have an equivalent structure known as the melon) as well as an oil-filled spermaceti organ. These massive structures aid in the transmission of sound for echolocation, and internal reinforcement of the junk allows it to be used as a battering ram. Sperm whales are not the only toothed whales to use their heads aggressively: orcas use violent motions of their heads, as well as their tails, to stun prey before drowning it. Might ancient predatory sperm whales have weaponized their foreheads in the same way?

Livyatan was not the sole arch-predator of Miocene Peruvian seas. It shared its habitat with another giant carnivore, the enigmatic megatoothed shark *Otodus megalodon*—better known simply as "Megalodon." The popularity of this famous shark is disproportionate to our understanding of it. Represented entirely by teeth and the occasional vertebra, much about this animal—its size and proportions, its relationships to other sharks, and even its appropriate scientific name—remains controversial. It may have been up to 15 m long and similar to a great white shark in behavior, and it may have competed with *Livyatan* for prey. But take all of this with a grain of salt: the decaying Megalodon jaws in this painting represent more anatomy of this shark than has ever been found as a single fossil.

The Aquatic Sloth, Thalassocnus *(Miocene–Pliocene)*

IT WOULD BE DIFFICULT FOR FOSSIL SLOTHS TO PRESENT A GREAT-er contrast with their extant, famously slow tree-living descendants. Modern sloths spend much of their time hanging from canopies of South America rainforests with strongly hooked claws, occasionally moving around to browse leafy vegetation. Their low-speed physiology probably reflects an energy-poor diet of hard-to-digest foliage, although some species supplement this with more nutritious food, such as insects and fruits. Sloths have many biological quirks, such as their weekly trips to the ground to defecate, their surprisingly strong swimming abilities, and their cultivation of camouflaging algae in their fur. Further peculiarities lay in their evolutionary history.

Sloths—also known as folivorans—are part of the South American mammal lineage Xenarthra, an anatomically radical group that also includes armadillos, anteaters, and the extinct, ankylosaur-like glyptodonts. Sloths appeared in the early Eocene and achieved a much broader diversity of size, geography, and ecology than is suggested by their living representatives. They became elephant-sized armored megaherbivores, bear-sized burrowers that excavated their own caves, and even aquatic species that foraged in shallow marine habitats. Sloths and other xenarthrans colonized North America when volcanic eruptions in Panama created a land bridge to South America in the latest Pliocene. This connection ended sixty million years of isolation for South America, and it allowed the faunas of the different landmasses to interact for the first time. This event, which transformed the biotas of both continents, is known as the "Great American Interchange."

The extinct giant sloths are characterized by huge claws on all limbs, necessitating many species to walk on the knuckles of their hands and the sides of their feet. Their claws likely had roles in defense, digging, and bringing foliage toward the body, whereupon powerful, muscular

lips could bring leaves into the mouth. A short, stout tail may have propped up rearing sloths when they stood on two legs to feed. Trackways suggest that large sloths walked on all fours, though the bulk of their weight was carried by their hind limbs. Excellent sloth fossils exist in both Americas, showing that the last of the giant and bear-sized sloths existed just eleven thousand years ago—recently enough that mummified remains, including fur and feces, have survived to modern times in caves.

One of the most remarkable fossil sloths was *Thalassocnus*, a genus comprising five species adapted to aquatic life. This realm was never explored by other xenarthrans, making *Thalassocnus* a pioneer for its group. Over time, *Thalassocnus* species became increasingly adept at swimming and foraging in marine habitats. The oldest Miocene forms were semiaquatic animals that, judging from their tooth wear and bone chemistry, ate vegetation adjacent to beaches, while Pliocene species show evidence of entering deeper water to eat seagrasses and algae. Along with increasingly long snouts and stronger lips (indicated by enlarged facial nerve openings), these more aquatically adapted *Thalassocnus* also had dense bones that acted as ballast in water, compensating for the buoyancy created by air in their lungs. *Thalassocnus* lacks strong adaptations for swimming and was likely a bottom-walker, using its large claws to grip the seafloor. A slight broadening of its shins and forearms may have enhanced their performance as paddles however, and the hind-limb reduction seen in more aquatically adapted species hints at increasing commitments to swimming behaviors in the last of the *Thalassocnus* lineage. We might speculate that, had they not gone extinct, another few million years of evolution might have shaped *Thalassocnus* into species resembling dugongs or manatees in appearance and lifestyle, sloths entirely adapted to aquatic life.

Grasses, Near-Hyenas, and Horses (Miocene)

GRASS IS SUCH A UBIQUITOUS PLANT IN THE MODERN DAY THAT it's difficult to imagine a world without it. In fact, though grasses and sedges have their origins in the Cretaceous Period, large grassy areas such as prairies, savannas, and steppes did not appear until the Oligocene and Miocene, when woodland environments gave way to open spaces. A shift to cooler and drier global climates probably played a major role in this change, along with greater seasonality and increased wildfire frequency favoring resilient, opportunistic, and rapidly growing grasses over trees and shrubs. Grasses have long been grazed by herbivorous animals (fossilized feces show that grasses were even part of dinosaur diets), but the establishment of huge grasslands allowed mammals to become specialized graminivores—species that eat little else but grass. As tough, abrasion resistant plants, grasses are difficult to digest, forcing graminivores to evolve sophisticated digestive systems to extract their nutrients. Among the greatest challenges of eating grass are tiny silica crystals known as phytoliths. These exist inside grass leaves and rapidly wear down the teeth of animals that chew them. Grazing mammals have responded to these with stronger, deeply rooted teeth that allow them to crush and grind grasses in powerfully muscled mouths. In turn, rather than defending themselves with toxins or thorns, grasses have adapted to cope with graminivores through an elevated regenerative capability, their leaves growing from the base of the plant and thus continually regenerating after they are eaten.

Unlike woodlands or scrubby forests, grasslands are exposed settings that offer little opportunity to hide from predators, and good grazing pastures can be situated great distances from each another. Mammals exploiting these new habitats were forced to adapt to face these challenges. Horses, a perissodactyl clade with origins in the early Eocene, became grassland specialists in the Oligocene and Miocene after spending most of the earlier history as woodland animals. Many of their adaptations to grassland life are typical of other mammals living in the same open habitats. Their large body size increased their travel efficiency and deterred predators, while relocating their eyes toward the top and back of their heads allowed them to scan for danger even when feeding. Elongation of limb bones below the knee and elbow, and reduction of toe and finger counts, adapted horse limbs for fast running, a useful trait for traveling long distances as well as for evading predators. *Hipparion*, a common Miocene–Pleistocene pony-sized horse of northern continents, embodies these features. Shown here, it likely resembled a modern horse in most regards, although if we saw one we would immediately note its three-toed feet. *Hipparion* walked on single hoofs on each limb, but two small toes were situated on either side of their principle digits. Accumulations of *Hipparion* fossils suggest it lived in groups or herds, a common antipredator behavior among modern grassland mammals. Digestive specialisms of perissodactyls mean that *Hipparion* was probably better equipped for eating dry, high-fiber foliage than the artiodactyls it coexisted with, allowing vast herds of these differently adapted mammals to coexist on the same grasslands.

The presence of many herbivore species in grassland settings allowed carnivorous mammals to develop their own plains specialists, adapting their anatomy to hunt big game in these new environments. Fossils show that larger members of the Carnivora (the guild that includes most mammalian carnivores) principally exploited these niches. Dogs, cats, bears, and their extinct relatives either adapted to endurance running to chase down prey, or developed stealthy behavior and camouflage coloration to ambush unwary animals. One of the most formidable Miocene carnivorans was the lion-sized *Dinocrocuta gigantea*, shown here. This large (perhaps 200 kg) predator was equipped with an enormous, powerful skull and bone-smashing teeth, and was thus very reminiscent of a hyena. *Dinocrocuta* was accordingly once thought to be a member of the hyena clade, but it is now considered a percrocutid, a close relative of this group. Healed injuries in the skull of a hornless rhinocerotid, *Chilotherium wimani*, match the teeth of *Dinocrocuta* and show that it pursued live animals. If percrocutids hunted like the largest living hyaenids (the spotted hyena, *Crocuta crocuta*), prey would have been worn down over long chases and weakened from crippling bites to the legs and abdomen. We might imagine that, as with most mammalian carnivores, *Dinocrocuta* would have begun eating their prey as soon as it was too exhausted or injured to run further, regardless of whether it was still alive.

Pelagornis, *the Largest Flying Bird (Miocene)*

FLYING BIRDS ARE SO FAMILIAR TO US THAT IT'S EASY TO TAKE them for granted, but they are true marvels of evolution and adaptation. Nowhere is avian adaptability more obvious than in their flight styles. Though they all share essentially the same basic body plan, birds have shaped their proportions into a myriad of forms that permit flight styles for every type of habitat. Some species, like parrots and crows, are generalist fliers that are able to flap, glide, and maneuver with equal skill. Others, including many water birds, are reliant on steady, powerful flapping to fly long distances. Birds widely regarded as poor fliers, such as turkeys, pheasants, and other gamebirds, should actually be considered launch specialists, bursting into the air almost vertically before traversing several hundred meters to avoid danger. Hummingbirds have an almost insect-like flight mechanism where highly mobile, rapidly beating wings facilitate a steady and agile aerial capability.

The birds that make flight look easiest are those adapted for soaring, a flight mechanism that exploits uplifted air currents (deflected winds and thermals) and supreme long-distance gliding abilities to fly for extended periods without flapping. Soarers are characterized by long, narrow wings and relatively large body sizes. Modern birds with the largest wingspans (about 3 m), the wandering albatross and the Andean condor, are both soaring specialists. Fossils show that, in Deep Time, soaring birds were even larger. Members of the "pseudotoothed" bird lineage known as pelagornithids had the largest wingspan of any known flying bird at 6–7 m from wingtip to wingtip. Another bird, *Argentavis magnificens* (a teratorn, a group of predatory birds related to New World vultures), is sometimes reported as having an even larger, 8 m wingspan. This estimate is almost certainly too generous, however, as all known *Argentavis* remains are smaller than those of large pelagornithids: a wingspan of 5–6 m is more likely. Pelagornithid proportions recall an exaggerated version of the albatross body plan and their wing spread may have approached the maximum size for a flying bird, any further wingspan increase being curbed by the hind-limb-dominant takeoff strategy common to all avians. Despite their magnitude, the largest

pelagornithid (*Pelagornis sandersi*) probably weighed just 20–40 kg, thanks to its small body and short hind-limbs. It was essentially a beak with a pair of giant wings attached.

Comparative anatomy and flight models show that pelagornithids were supreme marine soarers. They might have ridden on strong winds like albatross, moving over waves using only minute motions of their wings to control their passage, or perhaps soared to great heights on over-sea thermals like frigate birds. They seem to have had restricted abilities to flap, and they probably limited this action largely to take off. The cosmopolitan distribution of pelagornithid fossils indicates an ability to cover long distances with ease. We cannot know for certain how long these birds could remain airborne, but if they were like modern oceanic soarers, they could have spent most of the year flying around the planet, returning to land only to lay eggs and feed their offspring. Their ferocious-looking jaws were equipped with toothlike spikes along each edge: these "pseudoteeth" recall the dentition of fish- or squid-eating animals, and pelagornithids surely foraged for these creatures as they toured oceans and seas. A highly modified shoulder-wing joint makes their ability to take off from water questionable, and it's possible that they caught most or all their prey while in flight.

Pelagornithids are an ancient bird group, first appearing in the Eocene. Their relationship to other birds has been a matter of debate as, until recently, well-preserved and complete pelagornithid remains were unknown and comparisons to other bird species were limited. Traditionally, pelagornithids have been allied to other large marine bird groups, such as pelicans or albatross, but newly discovered fossils hint at affinities to nonsoaring birds including game birds (chickens, grouse, and pheasants) and waterfowl (ducks and geese), though these ideas are not without critics. While neither gamebirds nor waterfowl are known for their soaring flight, they had wide adaptive range in the early Cenozoic, including wader-like forms and giant flightless herbivores like *Gastornis* and mihirungs. Is it inconceivable that this lineage could develop giant soaring forms as well?

Gigantopithecus *(Pliocene)*

OUR INTELLECTUAL BIAS TOWARD HUMANITY'S OWN EVOLUTIONary history means that the diversity of nonhuman fossil apes is often overlooked when we recount the story of life. Humans and other modern apes—gibbons, orangutans, gorillas, and chimpanzees—arose as part of an evolutionary radiation of primates that began in the Miocene and saw apes of many kinds spread across Africa, Europe, and Asia. Ape fossils are rare but, largely through accumulations of fossil teeth and jaws in caves and other sheltered settings, we know that many parts of the world were once inhabited by multiple, coexisting ape species. The fossils of our own *Homo* line are among these, and scientists are still assessing how we—a lineage of technologically advanced, highly adaptable ground apes—fit into these communities, and when we began to significantly influence the history of our relatives.

Among the most mysterious of all fossil apes is *Gigantopithecus blacki*, a large species known from Miocene–Pleistocene fossils of southeast Asia. Thousands of teeth of *Gigantopithecus* have been recovered, but the rest of its skeleton, save for a handful of broken lower jawbones, is entirely unknown. This leaves much about this famous primate shrouded in mystery, and any reconstruction of it—including the one opposite—is extremely speculative. Even its size is not well constrained. Its teeth and jaws are slightly larger than those of the biggest living ape, the gorilla, and it is generally assumed that *G. blacki* was among the largest apes of all time. However, with only a few broken jaws hinting at the size of the skull, and no idea how large the head was in relation to the body, our size estimates are wide-ranging and of questionable reliability. Conservative estimates suggest that *G. blacki* may have stood just a little taller than a large gorilla (around 2 m), while others predict a gigantic standing height of 4 m. The latter seems overoptimistic given the size of the jaw fossils, and the depiction opposite accordingly shows an animal somewhat, though not unduly, bigger than a silverback gorilla. It towers over the early human (*Homo erectus*) in this scene, but chiefly because *H. erectus* is somewhat smaller than *H. sapiens*, with an average height of about 1.65 m.

Gigantopithecus is thought to be a member of the orangutan line (Ponginae), but it is unlikely to have been a giant version of these living apes. At times *Gigantopithecus* has been restored as a fully upright, somewhat humanlike ape because features of its lower jaw have been linked to a vertical neck posture. This idea has caught on in some circles, particularly among cryptozoologists hoping that Sasquatch or yetis might be surviving *Gigantopithecus*, but in reality this interpretation is very speculative: there is nothing about *Gigantopithecus* fossils that convincingly indicates an upright posture. If it was anything like large living apes, *Gigantopithecus* was likely a quadruped, and, in being the size of a big gorilla, it probably did not spend much time in trees. Its jawbones imply a relatively short, deep skull, and extensive wear on its teeth suggest the presence of very large, powerful jaw muscles, perhaps more akin to those of gorillas than orangutans or other apes. The same dental wear patterns imply a diet of very tough, coarse vegetation. Bamboo was abundant in the tropical regions inhabited by *Gigantopithecus* and was likely a common food source, along with other types of foliage. Such a diet implies a large gut to digest fibrous plant matter, another feature that adds to our mental image of *Gigantopithecus* as a heavyset, ground-based herbivore. These lifestyle inferences contrast with the ecology of orangutans, which mainly forage for fruit and insects in trees. Perhaps, despite its pongine affinities, *Gigantopithecus* was much more gorilla-like than orangutan-like in habit and form. These ideas are mainly conjecture, of course, and will remain so until we develop a better understanding of *Gigantopithecus* anatomy.

Insect Societies and the Giant Asian Pangolin (Pleistocene)

MANY ANIMAL SPECIES EXPLOIT THE ADVANTAGES OF LIVING IN groups, but few have made sociality and cooperation as essential to their existence as the eusocial insects. These are insect species with highly organized societies, characterized by a cooperative approach to rearing young, the year-round presence of adults, and segregation of individuals into castes (most often into roles for reproduction [queens], foragers and builders, guards, and flying individuals that disperse and create colonies elsewhere). With all members of the colony sharing the same genes, eusocial species are considered "super organisms": species for which life as individuals is impossible, and only collectives can survive.

The hymenopterans (bees, ants, and wasps) and termites are the most dedicated eusocial insects. Ants and termites are major contributors to Earth's biomass thanks to some species forming colonies with millions of individuals. The nests constructed by these animals rank among the most sophisticated natural structures on the planet, drawing parallels to human settlements in their ability to provide their inhabitants with comfort and security. Nests can be made of many materials in varied settings, including the inside of rotting wood, within soil, as mud mounds of varying size (some gigantic, being many tens of cubic meters in volume), or constructed in elevated positions (such as within tree branches or under overhanging rock) using paper or wax. Each nest provides its owners with safety and shelter, as well as dedicated spaces to store or farm food and to raise offspring. Particularly sophisticated nests, such as termite mounds, have mechanisms to keep the inhabitants cool against elevated external air temperatures.

The patchy insect fossil record means some uncertainty exists regarding when various insect lineages committed to communal living, and fossils of ancient nests are very rare. Many alleged fossils of Mesozoic nests have been identified, but most of these lack characteristic features of true insect colonies and their identification as ancient ant or termite structures is highly controversial. Recently discovered fossils of Early Cretaceous termites preserved in amber show that they had evolved eusocial behavior by one hundred million years ago. Ants of this time probably also lived in groups, though perhaps not yet in huge colonies. The capability to farm fungus for food is predicted to have appeared at some point in the Cenozoic for both groups, an idea consistent with the identification of fossil "fungus gardens" among Tanzanian Oligocene termite nests. Evolutionary models predict that Late Cretaceous bees had also developed various grades of social behavior.

Insect colonies represent enormous quantities of protein to any animal that can breach their nest defenses to harvest their inhabitants. To ants and termites, such creatures are the stuff of nightmares: animals with huge claws and powerful limbs that can excavate their underground shelters or smash through walls; thick, sometimes armored skin that resists attack from guard castes; and extremely long, sticky tongues that extend through nest corridors and chambers to grab panicked citizens. These are creatures like anteaters, armadillos, aardvarks, and—as shown here—pangolins. Though sharing similar adaptations, these animals are not closely related, their common features being the result of convergent evolution. Pangolins seem to have evolved in the Eocene from the same branch of mammals that gave rise to Carnivora, though their slow, trundling habits and massive scales (made from the same material as fingernails) clearly distinguish them from their carnivorous cousins. Today confined to Africa and Asia, pangolins also once roamed Europe and North America. A giant Asian species—the 2–2.5-m-long *Manis paleojavanica*—lived in parts of Indonesia during the Pleistocene. Much of southeast Asia was covered in savanna-like habitat at this time and was populated by mound-building *Macrotermes* termites: ideal prey for this large pangolin. The extinction of the giant Asian pangolin coincides with the arrival of humans into its range, and human predation may have had a role in their demise. Similar fates await the eight pangolin species alive today, all of which are hunted in huge numbers because of pseudoscientific beliefs about the medicinal properties of their scales, as well as for their meat. The long-lived pangolin lineage may be extinct before the year 2050 because of these practices.

Woolly Mammoths (Pleistocene)

FEW SPECIES ARE AS EMBLEMATIC OF OUR RECENT ICE AGE AS *Mammuthus primigenius*, the woolly mammoth. The recovery of numerous frozen mammoths from permafrost in Russia and Alaska, as well as abundant skeletal remains from other countries, has provided an extremely detailed insight into their biology. Our mammoth specimen inventories span the full spectrum of age and gender and preserve their tissues to the cellular level, allowing for exceptional insights into their genetics, stomach contents, life appearance, and growth regimes. These large proboscideans are closely related to Asian elephants and existed in large numbers across the Northern Hemisphere throughout the Pleistocene. They were among the last of the mammoths, but they were not, as sometimes depicted, the largest—they were actually similar in size to modern African bush elephants. They were well adapted to life in cold climates by having small ears, a short tail, sophisticated body integument comprising dense underfur and longer outer hair, and a layer of insulating fat around the body. Stomach contents show that they did not live among deep snow and glaciers but spent much of their time in open habitat known as "mammoth steppe"—a cold-adapted grassy ecosystem that still survives in a handful of locations in Siberia. Mammoths and other large grazers were essential to maintaining these environments, their removal of shrubs and trees preventing forests from overwhelming the grasslands.

Woolly mammoths have a long and complex history with several types of humans, including early members of our own species and *Homo neanderthalensis*—the Neanderthals (opposite). We both relied on mammoth bones and tusks for a variety of uses, including the fashioning of tools as well as the creation of shelters. Human hunters would have wanted to remain on mammoth steppe because of the wealth of game but, without forests, wood would have been scarce, and mammoth bones were often the only objects available to create large structures. Multiple examples of humans having processed mammoth carcasses are known, but how frequently early humans hunted mammoths remains controversial. Healed spear wounds in mammoth fossils record failed predation efforts, but they are not particularly common. Mammoths were surely dangerous game for our Pleistocene ancestors, so perhaps even powerful, robust humans like Neanderthals generally avoided them in favor of smaller, less challenging prey. However frequently we hunted them, mammoths clearly made an impression on early humans, their forms being depicted frequently in Pleistocene cave art.

Our relationship with and reverence for woolly mammoths has not halted despite their extinction fourteen thousand to ten thousand years ago. We continue to use their tusks to create art and objects, in part because the decline of living elephants and ever tightening antipoaching laws make it easier to trade ivory from mammoths than from modern proboscideans. The resurrection of mammoths by cloning is also a continued point of discussion among scientists and journalists, a question with more at stake than simple scientific curiosity: if a long-extinct animal such as a mammoth can be re-created, extinction need not be forever. Opinion is split about the feasibility of cloning mammoths, though even optimistic scientists would concede that numerous hurdles currently stand in the way of this goal. Even the best-preserved mammoth genetic material is degraded to an extent that it is not a viable blueprint for a living individual, and—even if we had a perfect sample—the complexities associated with turning a genome into a flawless set of chromosomes, and ultimately mammoth cells, are huge. If we get to the stage of having a lab-grown mammoth egg cell, we then face the challenges of implanting them into surrogate mothers—2.5 tonne Asian elephants. These animals are rare, are protected because of their endangered status, are an inevitable practical nightmare as laboratory subjects (consider that we'd need many, many elephants to have a shot at cloning success), and they're also prone to developing tumors when their reproductive cycles are interrupted. Science aside, this latter point is the tip of an ethical iceberg for this most ambitious "de-extinction" project. In all likelihood, mammoths will remain extinct, at least for the foreseeable future.

A Dwarfed Giant Horned Turtle (Pleistocene)

WE HAVE MADE IT NEARLY TO THE MODERN DAY WITHOUT MEET-ing one of the most familiar, and also most bizarre and mysterious, of all reptiles: the testudinatans—better known as turtles. The turtle branch of evolution is a long one, stretching back to at least the Triassic, with a Permian origin predicted in some evolutionary models. Although oc-cupying a range of habitats, body sizes, and lifestyles, throughout that time turtles have remained dedicated to their Bauplan of a shelled body; a beaked face; and short, stout limbs. The number of living turtle species is estimated to exceed 350, but some data indicates that this count may be too conservative, and that as many as 470 species may exist today. At least half of these are at risk of extinction, and a full third are classed as "endangered" or "critically endangered," mostly because of habitat degradation.

The relationships of turtles with other reptiles are hotly debated. Turtles lack the twin jaw muscle openings that characterize most reptile skulls, and this has linked them with the Parareptilia, a group that con-tains the pareiasaurs and procolophonids we met earlier in the Permian and Triassic Periods. Parareptiles branched off the reptilian tree close to its roots, and if turtles belong to this group, they would be the sole survi-vors of a very ancient reptile lineage. Turtle DNA, however, suggests that this is incorrect. Genetically speaking, turtles seem more closely related to modern reptiles, either as close relatives of lizards and their kin, or as cousins of the archosaurs. Uncertainty on this issue is compounded by a historic lack of fossils representing the early phases of turtle evolution. For many decades, the oldest known member of the turtle line was the Triassic *Proganochelys quenstedti*, a relatively "primitive" turtle com-pared to later species, but still a fully formed turtle with all the unique anatomical features of the group. Thankfully, newfound Chinese fossils have begun to shed light on the morphology of turtle ancestors and, although their evolutionary significance is contested at present, their discovery gives hope that we will eventually pin down how turtles are related to other reptiles.

Turtles are anatomically alarming even to experienced biologists. Though outwardly unassuming, they are some of the most extremely modified of all tetrapods. Their shells are made up of two elements: an underside (plastron) and an upper portion (carapace). These are joined by bony bridging elements running along the side of the animal. The carapace is made from bone formed within the skin, as well as incorpo-rating ribs, vertebrae, shoulder blades, and pelvic elements. It requires some serious rejigging of typical tetrapod anatomy to enclose the limb girdles within the rib cage! Virtually all turtle-line reptiles have tooth-less, beaked mouths, but only some species can withdraw their heads and necks into the shell. Most turtles, including a large portion of living species, can pull their head and neck in only sideways, partially covering those body parts with the front of the shell. Turtles likely had a terres-trial, burrowing origin, but they have adapted to aquatic life numerous times in their evolutionary history. It's for this reason that terms like *tortoise* and *terrapin* lack strict definitions: these are categories for life-style and appearance, not natural evolutionary groupings.

Oceania was once home to a remarkable turtle group: the meiolani-ids, or horned turtles. Meiolaniid origins stretch back into the Creta-ceous, and these animals must have been remarkable to see alive, being large (up to 2.5 m long) and bearing spiked, armored tails in addition to impressive-looking cranial horns. They were fully terrestrial in habit and probably subsisted on low-lying vegetation, primarily grasses. They became extinct only three thousand years ago. Among the last of their kind was the 1-m-long dwarf island species, *Meiolania platyceps* (shown here), which existed on Australia's Lord Howe Island. Fossil sites with hundreds of human-butchered *Meiolania* bones suggest that human hunting had some role in their extinction.

Moa (Holocene)

FOR THE LAST EIGHTY MILLION YEARS, NEW ZEALAND HAS BEEN isolated from the other southern continents, allowing for the development of unique ecosystems containing many types of flightless bird. Flightless parrots, rails, and kiwis survive today, albeit in such low numbers that many species only survive through intensive conservation efforts. The largest and most spectacular of New Zealand's flightless birds, the moas, are no longer with us, following zealous predation by human settlers arriving in New Zealand just a few hundred years ago. Moas are ratites, and thus part of the same avian lineage as ostriches, emus, and kiwis, though they are not closely related to kiwis despite their shared New Zealand home. DNA analysis has revealed the surprisingly complicated evolutionary history of ratites, wherein moas have closer affinity to the South American tinamous than they have to kiwis or even Australian ratites, such as cassowaries and emus. This is far from the only shakeup of bird relationships brought on by genetic data. Many branches of the avian family tree are now differently arranged compared to evolutionary relationships deduced from physical anatomy alone.

Except for a few highly stylized paintings by historic Maoris, no records detailing moa appearance or habits are known. Thus, while moas died out around 1400 CE—a nanosecond ago in the context of geological time—we must reconstruct their biology as if they had been extinct for millions of years. This is a sharp reminder of the finality of extinction, and a sobering thought given the critically low populations of so many species across the planet today. The moa fossil record begins 16-19 million years ago, but it took just 150 years from the arrival of humans in New Zealand to reduce their populations to unsustainable levels. Archaeological sites preserve the rafts and cooking tools which transported and processed the carcasses of these often-enormous birds, animals that—until our arrival—had never been concerned about large ground predators. It seems the only natural predator of adult moa was the Haast eagle (*Harpagornis moorei*), a predatory bird comparable in size to our largest living raptors. As moa populations dwindled, the Haast eagle also became extinct. Both are part of one of the largest extinction events of the Holocene Epoch: the loss of thousands of bird populations throughout Polynesia, brought on by humans spreading south.

Moa have left behind an extensive fossil record including skeletons, mummified remains, and footprints, providing a detailed picture of their biology. Their modernity also means we have access to their genetic information, and this has helped resolve a long-standing controversy over the number of moa species. Moas vary in many aspects of proportion and body size, and we once had Pleistocene and Holocene moa divided into over twenty species. Only nine are now recognized, however, thanks to DNA showing that pronounced size differences between moa individuals were a result of gender dimorphism, not taxonomic separation. Female moas were typically the larger gender, with female members of the *Dinornis* genus being particularly huge: 280 percent larger than males. This is the most significant dimorphism of any bird species or land mammal.

Though superficially ostrich-like in appearance, moas were generally heavyset birds better adapted to walking than running. Some, like the giant moa (*Dinornis robustus*) shown here, were huge animals that stood taller than a human and weighed almost a quarter of a tonne, but others—like the bush moa (*Anomalopteryx didiformis*)—were much smaller, just over 1 m tall and only 20–50 kg in mass. All had lost their wings entirely, the only hint of a forelimb being a finger-sized remnant of their shoulder girdle. At least some were covered head to toe in feathers, though whether this was true of all species remains unknown. Their diets and preferred habitats have been revealed by preserved stomach content and careful assessments of their fossil sites. All moas were herbivores, but they differed in their habitat and food choices, allowing many species to live across the varied landscapes of New Zealand without stepping on each other's ecological toes.

Modern Humans (Holocene)

APPROXIMATELY THREE HUNDRED THOUSAND YEARS AGO MODern humans, *Homo sapiens*, speciated from other members of the *Homo* group in Africa. Genetic data shows that we are a mongrel species, having interbred with other *Homo* species encountered in our travels away from our African home. The DNA of Neanderthals and another, largely mysterious human lineage, the Denisovans, still lingers in our genes.

Humans are part of the ape lineage Homininae, a group of especially intelligent primates that, on the evolutionary path to you and me, became especially adept at technological innovation, problem-solving, and communication. Though seemingly anatomically odd compared to other apes—most obviously because of our long legs and short arms (these seeming to be adaptations to an obligate upright stance and long-distance travel), our anatomy is entirely normal for primates. Our limb proportions, for example, are not so different from those of monkeys. Our bodies look largely devoid of fur, but only because our hair is mostly a short, fine pelt that gives us the appearance of naked skin. We are, in fact, as hairy as any other primate. Our adaptations for standing and walking on two legs are better developed than those of our relatives, but are not unique to us either: many primates have adaptations for bipedality. Our faces and bodies are also ornamented with hair and fatty tissues to advertise our vigor and fitness, just like other primates. And beyond our anatomy, even our social structure, though culturally varied, is primate-typical, revolving around prolonged care of offspring and the formation of clans comprising related and unrelated individuals.

What really sets us apart from other apes, and perhaps all other animal species, is our technological prowess. Tool use is common to animals of all kinds, even in species that lack grasping hands, such as birds and dolphins. But no other species have developed their technological capability to a point where they can bypass many major challenges of natural selection. Our technology has removed barriers against our spread and proliferation, allowing us to quickly overcome natural obstacles that would take typical evolutionary mechanisms multiple generations to respond to, and our unprecedented ability to store collective knowledge allows successive generations to improve on the innovations of our ancestors. These abilities have allowed us to transform much of the planet into settlements entirely conducive to our own needs and safety, and in the process we have created a new, now-widespread form of habitat: the urban environment. Humans are an exceptional and remarkable lineage, the likes of which has never existed on Earth before.

Our success has not been without cost, however. The results of our continued population expansion and resource use have come into sharp focus in recent years, it being apparent that human activities are having environmental impacts on a global scale. The emissions from our chosen energy sources are causing rapid, worrying shifts in global climate. Our waste and pollutants are found in all habitats and environments around the world, even drifting into the deep ocean and to landmasses uninhabited by humans. Widespread transformation of land into farms and homes has left little wilderness and has reduced animal and plant populations to dangerously low levels. Our actions are now even detectable at the geological level, leading some to propose an "Anthropocene" Epoch—a period of geological time defined by the presence of humanity.

These actions are causing a biodiversity crisis comparable to the "Big Five" great mass extinctions. We are living through a biological cataclysm, and our data is clear: we are the cause. Long-term degradation of the biosphere means that uncountable numbers of species are at risk, whereby even relatively common "low concern" plants and animals are struggling. This crisis will ultimately affect us too. Widespread waste and pollution, collapsing marine and terrestrial ecosystems, and shifting climates are already impacting food availability and quality, our health, and the habitability of our towns and cities. The impact we are having on Earth is not just a problem for wildlife: it is a problem for *all life*. How we choose to respond to these facts in the next few years and decades will determine as much about our own future as it will the fate of the natural world.

Nene (Holocene)

IN 1778 CE, THE BRITISH EXPLORER CAPTAIN JAMES COOK LANDED the HMS *Resolution* on the Hawaiian Islands. The arrival of Europeans on Hawaii instigated major changes to the natural history of the islands, bringing a suite of human hunters and habitat alteration as well as introducing cats, mongooses, and pigs into ecosystems ill-equipped to cope with them. Hawaii's wildlife had suffered a suite of extinctions since the arrival of Polynesian settlers in the ninth to tenth centuries, but the influx of Europeans increased the extinction rate considerably. Among the affected species was the Hawaiian goose, or nene (*Branta sandvicensis*). This mid-sized, soft-voiced goose is endemic to Hawaii, having evolved from Canada geese blown to the islands in storms. It is characterized by half-webbed feet, long legs, distinctively textured plumage, and relatively terrestrial habits compared to other geese. From an estimated population of about twenty-five thousand in 1778, populations fell to just twenty to thirty wild birds in 1951, with thirteen birds in captivity. It seemed near certain that the nene would join the hundreds of endemic Hawaiian animals to have become extinct in the last thousand years.

Nene, however, are still with us, and their wild population is slowly growing. They are alive today because the same species that almost drove them to extinction decided to save them. From the 1950s onward, intensive nene conservation efforts were introduced that included captive breeding and release programs, control of introduced predators on Hawaii, and designation of sanctuary areas for wild goose populations. From those twenty to thirty birds, over one thousand wild Hawaiian geese now exist, with another thousand in captivity in zoos and wildlife parks around the world. The species is still at risk because of its small population and considerably reduced genetic diversity, but it now has a fighting chance of survival.

The story of the nene has been replicated numerous times by biologists and conservationists across the world. Restorative efforts have seen iconic species like pandas become abundant enough in the wild to be removed from endangered species lists. Eradication of introduced species on islands and careful control of fishing quotas are restoring ecosystems to historic balances and are allowing native species to re-establish themselves. Entire populations of endangered amphibians have been captured to exist in captivity to preserve them against habitat loss. The amount of effort that goes into such schemes is incredible: even small conservation projects have enormous practical and organizational demands, as well tremendous scientific and economic requirements. The challenges of rearing breeding populations or keeping tabs on precious wild individuals are vast, and can seem futile against the continued onslaught of habitat degradation, poaching, and shifting climates. In reality, the resources needed to preserve species and environments often fall well short of what is needed, and not all efforts are successful. The sad fact of modern conservation is that we have to choose our battles, weighing available resources against conservation needs and their likelihood of success.

But at the core of any conservation effort is a decision that a species or an environment is worth saving, and with enough groundswell action can happen. Our biodiversity and environment are at crisis point, and it is only through changing our views on the importance of the natural world, and taking responsibility for the impact we've already had, that we will avert further loss among our remaining species and habitats. Unlike the meteorite strikes or volcanism that catalyzed previous extinction events, we are a mass extinction with a conscience. For the first time in our planet's history, the decisions of conscious, sentient beings will determine the shape of life through future ages.

Appendix

Artist's Notes

I HAVE MADE EVERY EFFORT TO MAKE THE PAINTINGS IN THIS book as credible to contemporary science as possible, but they remain a blend of hard data, supposition, and speculation. The ratio of these elements varies from image to image, and their proportion may not be directly apparent to most readers. In the interest of making the paleoart in this book as scientifically transparent as possible, overviews of the basic data and ideas informing my paintings are given below. If more detail about the methods and philosophy of reconstructing fossil animals used herein is desired, it can be found in *The Palaeoartist's Handbook* (Witton 2018). I have also used this appendix to note some of my compositional ideas, and to indicate where some of the works presented here nod to the original *Life through the Ages* (hereafter *LTTA*)—in order to give Knight due credit for inspiration and subject choice.

BUILDING A PLANET FIT FOR LIFE (HADEAN)

Though some aspects about the formation of our planet are still debated, the main tenets of this image—a superheated proto-Earth, a ring of dust and gas, abundant asteroids—are not controversial. Many of these same planet-forming phenomena are now being directly witnessed in association with distant, newly forming worlds: see Keppler et al. (2018) and Müller et al. (2018) for modern insights into planetary formation. Our own Earth must have looked something like the form illustrated in this image, at least during its earliest years of existence.

THE ORIGINS OF LIFE (ARCHAEAN)

This murky scene was based on photographs of deep-water hydrothermal chimneys, specifically those of the Lost City Hydrothermal Field in the north Atlantic. These environments are so deep that they escape all light, but pitch dark conditions would make for a lousy picture. Dim illumination was added to show the cityscape-like features of these alien settings.

STROMATOLITES (ARCHAEAN–PROTEROZOIC)

Stromatolites are rarely given any dignity in paleoart: for some reason, aggregates of mud, slime, and rock just don't inspire artists very much. Reconstructing this image, informed mostly by photographs of living stromatolites, was entirely about giving a stromatolite the sort of artistic treatment usually reserved for *Tyrannosaurus rex*: atmospheric, imposing, and completely overblown.

THE EDIACARAN BIOTA

Though each Ediacaran animal in this scene is restored based on contemporary ideas of lifestyle and appearance, much about their paleobiology remains controversial, so please appreciate my reconstructions of these species with the same skepticism that should be applied to all Ediacaran paleoart. The especially mobile and curious-looking *Kimberella* is largely based on interpretations presented by Ivantsov (2009). Artists often compose Ediacaran scenes as if they were creating a shop display, arranging them in a neat, uncluttered fashion with each form clearly visible. I opted for a more complex, and I hope naturalistic, composition here, with animals draped over and moving between algae-covered complex topography.

THE CAMBRIAN BIOTA

Two major challenges face artists working on Cambrian animals. The first is the widely publicized ambiguity about the mode of life of some species; the other is the lesser-known issue of scale. Most fossils from sites such as the Burgess Shale—which forms the basis of this painting—are of tiny animals, many of them just millimeters in length. Conversely, animals like *Anomalocaris* are much larger—nearly a meter long. The size differential is somewhat apparent here, but in order to stop the anomalocarids dominating the image, they are depicted somewhat smaller than their maximum size. The animals in this scene were mostly based on data presented by Erwin and Valentine (2013).

TRILOBITES (ORDOVICIAN)

Trilobite anatomy is well known but difficult to render in art because so many of their interesting features—their legs, gills, and antennae—are almost completely hidden beneath their carapace, and the use of lower points of view omits the characteristic anatomy of their dorsal surface. This may explain why artists often use a plain dorsolateral aspect in their paintings of these animals, a tradition I was unable to shake here. The *Ogyginus forteyi* and *Basilicus vidali* in this scene are based on early Ordovician fossils of Morocco (Corbacho and Vela 2010).

THE ORDOVICIAN–SILURIAN EXTINCTION

The soft tissues of the unhappy-looking orthocone in this scene are closely based on those of *Nautilus*, its closest living relative. Orthocones are much more closely related to *Nautilus* than ammonites, so while it is unknown if orthocones possessed structures like apertural hoods and eighty to ninety tentacles, basing their appearance on *Nautilus* is a relatively defensible—if still somewhat speculative—take on their life appearance.

JAWLESS AND JAWED FISH, AND "SEA SCORPIONS" (SILURIAN)

This late Silurian scene is based on data from the Downton Castle Sandstone Formation in Wales, a site that yields bone beds of acanthodian fossils but, somewhat less fortunately, often only as isolated scales. Accordingly, the fish and eurypterid depicted here are somewhat stereotyped examples of their respective groups, rather than detailed reconstructions of certain species. Märss, Turner, and Karatajūtė-Talimaa (2007) and Long (2010) were principle references for this scene.

PLANTS COLONIZE THE LAND (SILURIAN)

Mosses and other encrusting plants, rather than early vascular plants, dominate this scene. It seems likely that these organisms were more abundant than larger plants in early Paleozoic landscapes. Like the previous Silurian illustration, this painting is based on data from the Downton Castle Sandstone Formation, and the two images can be seen as land and coastal counterparts. Because this painting is set at the Welsh seaside, it is raining.

TITANICHTHYS (DEVONIAN)

Titanichthys is represented only by skull material, the rest of the skeleton being cartilaginous and thus having little preservation potential. The skull of *Titanichthys* is now relatively well documented, and its odd features—such as the downturned mandible—are verified by multiple specimens (Boyle and Ryan 2017). The only completely known placoderms are small fish that did not need to optimize their form for oceanic cruising, as necessitated by a large-bodied ram-feeding ecology. Inspired by arguments for a white shark–like form in the giant predatory placoderm *Dunkleosteus* (Ferrón, Martínez-Pérez, and Botella 2017), I gave *Titanichthys* a body shape more consistent with strong long-distance swimmers, reasoning that the same selection pressures that shape the bodies of modern fish would also have operated in the Devonian.

LAKES OF THE EARLY CARBONIFEROUS

This scene is based on new fossils from Early Carboniferous sites in Scotland, especially those from the Willie's Hole locality. These sites are, at time of writing, only provisionally explored and documented. The fauna shown in this scene is based on fossils from these sites but is necessarily somewhat anatomically stereotyped, pending further research into these deposits and their fossil assemblages. Its reconstruction was assisted by early tetrapod experts from National Museums Scotland and by the extensive library of skeletal diagrams provided by Carroll (2009).

PULMONOSCORPIUS, GIANT SCORPION OF ANCIENT SCOTLAND (CARBONIFEROUS)

Excellent, complete fossils of *Pulmonoscorpius* allow for detailed reconstructions of its life appearance: it really does look just like a giant version of a modern scorpion, despite its antiquity (see Jeram [1993] for an excellent overview of its anatomy). The small tetrapod it is chasing in this scene is *Westlothiana lizziae*, a lizard-like lepospondyl also known from excellent fossils—only portions of the tail are missing. Carroll (2009) was its principle anatomical reference. While the predator–prey scene show here is speculative, the small size of *Westlothiana* (20 cm adult length) means it may have been ideally sized prey for this giant scorpion.

THE TETRAPODS INVADE THE LAND (CARBONIFEROUS)

The animals in this scene, as per the Carboniferous lake illustration, are based on new Scottish discoveries that have yet to be researched in detail. The colosteid and whatcheeriids are modeled on well-understood species from these clades (see Carroll 2009), while the black and red animal to the far left—based on an unnamed specimen with the nickname "Ribbo"—remains only provisionally understood. Its reconstruction was assisted by early tetrapod experts from National Museums Scotland.

PLATYHYSTRIX, A SAIL-BACKED "AMPHIBIAN" (CARBONIFEROUS–PERMIAN)

The sail of *Platyhystrix* is essentially the only known part of this animal's anatomy, save for a smattering of vertebrae and a piece of skull. Thus, as with virtually all reconstructions of *Platyhystrix*, this reconstruction marries the body and skull of the dissophorid *Cacops aspidephorus* with the sail of *Platyhystrix*. What we know of *Platyhystrix* implies that this approach is a defensible one, but be mindful that more complete remains of this genus may eventually render such reconstructions obsolete. Proportional details for *Cacops* were largely taken from mounted skeletons in the Field Museum, Chicago.

CASEIDS: THE LAND VERTEBRATES DECLARE WAR ON PLANTS (PERMIAN)

Cotylorhynchus is represented by several spectacular fossils that are complete, articulated, and three-dimensional, leaving little doubt about the proportions of this animal. Indeed, if it weren't for the quality of *Cotylorhynchus* fossils, it might be difficult to believe that those tiny heads were attached to those massive bodies! These remains informed the principle anatomical reference of this image, with additional details taken from Stovall, Price, and Romer (1966).

DIMETRODON (PERMIAN)

This image is a direct update of Knight's *LTTA* original, a fairly conventional image of *Dimetrodon* snarling next to a stream. This update gives *Dimetrodon* more complex behavior (with feeding indicated by the

Diplocaulus carcass and parenting implied by the presence of juveniles) as well as a major anatomical overhaul. Recent work by paleontologist Scott Hartman has shown that our traditional idea of *Dimetrodon grandis* is erroneous in many areas: it likely had somewhat more mammalian spinal curvature, a shorter tail, and a more upright limb posture (Hartman 2016). The erect limbs are not reflected in this illustration because of the crouched pose of the animal, but it's otherwise an entirely twenty-first-century take on this classic paleoart subject.

HELICOPRION (PERMIAN)

Because so little is known about *Helicoprion* anatomy, I decided to focus entirely on its best-known feature: its jaws (Tapanila et al. 2013). The body is partly discernible but is shrouded in enough darkness to prevent its form being obvious. In addition, the extremely foreshortened view prevents easy calculation of its length. I am not a huge fan of the paleoart convention of hyper-foreshortening, but in this instance it helps to hide some substantial gaps in our understanding of this animal, as well as to create a slightly creepy image. That's also OK with me: many marine species look alarming when they're witnessed lunging out of darkness, and I'm sure the same was true in the past.

DICYNODONTS (PERMIAN)

Aulacephalodon peavoti is a common fossil in the Karoo Supergroup, though it remains known primarily from isolated skulls and bones of varying quality. Nevertheless, enough is known of this taxon to permit a reasonable reconstruction of its form. This situation also broadly describes our state of knowledge for the anatomy of *Cistecephalus microrhinus*. My *Aulacephalodon* reconstruction largely references a mounted skeleton at the Field Museum, Chicago, while Nasterlack, Canoville, and Chinsamy (2012) informed *Cistecephalus*. While my reconstruction is not a direct homage to *LTTA*, Knight featured an unnamed dicynodont from the Karoo Supergroup, making this image a continuation of his theme.

THE GREAT DYING: THE END-PERMIAN EXTINCTION

The animal in this scene is *Scutosaurus karpinskii*, an iconic fossil reptile from Russia that existed at the very end of the Permian. It's well studied and known from excellent fossil specimens, and can be reconstructed in detail: even the arrangement of the scales on its face and back are known. A mounted skeleton at the American Museum of Natural History provided the chief reference for this species. A trickier aspect to this image concerned visualizing the volcanism causing the end-Permian extinction event, given that no eruptions on Earth today come anywhere close in scale. I referenced aerial photographs of fissure eruptions to create the backdrop to this image, and regions undergoing long-term eruptive phases to create the foreground.

THE FOUNDATION OF THE MODERN AGE (TRIASSIC)

The landscape of this Brazilian scene reflects the lack of trees indicated by the early Triassic "coal gap" following the Permian extinction. The principle animal in this image, *Teyujagua paradoxa*, is known solely by a well-preserved skull (documented by Pinheiro et al. [2016]). The rest of its anatomy is based on other early archosauromorphs, particularly *Proterosuchus fergusi*. *Procolophon trigoniceps* is a better-known animal, with the bulk of its osteology represented by different specimens; its anatomy is detailed by DeBraga (2003). The temnospondyl in the background, *Sangaia lavinai*, is not well represented in fossils (only partial skulls are known), so it is presented here through some liberal referencing of close relatives.

AMPHIBIOUS ICHTHYOSAURS (TRIASSIC)

Most of *Cartorhynchus* is known from a crushed but well-preserved skeleton, the only missing portion being the bulk of the tail (Motani et al. 2015). However, a complete tail is known from the closely related (and possibly conspecific) ichthyosauriform *Sclerocormus parviceps* (Jiang et al. 2016), and it seems reasonable to assume that this is a good model for the *Cartorhynchus* caudal series. Of less certainty is the torso width of these early ichthyosaurs, as their crushed fossils are preserved only in lateral aspect. I have assumed a fairly rotund but narrow body—in line with a number of other Triassic ichthyosaurs, and also consistent with the rib curvature of the only known *Cartorhynchus* specimen.

REBELLATRIX (TRIASSIC)

Most of the *Rebellatrix* skeleton is documented by Wendruff and Wilson (2012), excepting the skull, which has yet to be found. This is somewhat unfortunate because we tend to focus on faces in art, even if they belong to animals. This required that I invent an appropriate cranium for this aberrant coelacanth. The heads of my *Rebellatrix* are based on other

fossil coelacanth crania but have been given a streamlined profile to correspond with their habits of powerful, rapid swimming.

ERYTHROSUCHIDS (TRIASSIC)

The species in this scene—*Garjainia madiba*—is not the best understood erythrosuchid, being represented by a number of broken bones from across the skeleton (Gower et al. 2014). Thankfully, other *Garjainia* species are more completely known, allowing for the *madiba* material to be mapped onto these species to build up a reasonable picture of its appearance. A mounted *Garjainia* in the Paleontological Institute of the Russian Academy of Sciences, Moscow, was a major reference for this painting (see Sennikov [2008] for images of this mount). One of the most distinctive features of this animal—the laterally prominent bosses above and below its eyes—meant it was important to have one animal facing the viewer in this scene.

ATOPODENTATUS, AN EARLY UNDERWATER HERBIVORE (TRIASSIC)

Complete and articulated skeletons of *Atopodentatus* are known, which allows for some confidence about its general proportions (Cheng et al. 2014; Chun et al. 2016). My application of ornamented skin, though speculative, is based on the sculpted backs of crocodylians and the spiny frills of marine iguanas: most marine animals are streamlined, but not all. The behavior depicted in this scene—a lonely *Atopodentatus* howling at a suspiciously shaped rock in the dead of night—is also entirely speculative. Reconstructing extinct animal behavior is extremely difficult, and it's likely that their habits—now lonely, vast, and far away, to our knowledge—were often more peculiar that we would typically imagine.

MORGANUCODON AND THE DAWN OF MAMMALS (TRIASSIC)

Though *Morganucodon* is known from lots of material, no complete specimen has been found and its precise proportions remain mysterious. Thus, this restoration is informed by *Morganucodon* anatomy as much as possible (e.g., Lautenschlager et al. 2017), but also owes something to *Megazostrodon*, a close relative known from more complete skeletons and illustrated in a classic F. A. Jenkins skeletal reconstruction (first published in Jenkins and Parrington [1976]). The fiery setting is based on burned plant material found in the same deposits as *Morganucodon*;

evidently, the island home of this animal was subjected periodically to forest fires.

THE GREAT CRINOID BARGES (JURASSIC)

The nature of the crinoid barges is—as outlined in the main text—well represented by fossils and is described at length in paleontological literature. The specimen referenced in this painting is the largest-known example of a crinoid barge, and, at 500 m² it is one of the largest fossil specimens in the world. It was recovered from the Posidonia Shale in southwestern Germany and occupies a huge wall display at the Urwelt Museum Hauff, in Holzmaden. The plesiosaurs in the scene are *Meyerasaurus victor*, a species known from an excellent, articulated skeleton from the same deposit. The amazing holotype specimen of this plesiosaur, on display at the Stuttgart State Museum of Natural History, formed the main reference for their anatomy. The small fish hovering around the barge are speculative, but they allude to the presence of noncrinoid organisms also using the logs as a habitat, a fact verified by close examination of the barge specimens.

OPHTHALMOSAURUS (JURASSIC)

Ichthyosaurs are often preserved as crushed specimens, but *Ophthalmosaurus* fossils include excellent and complete three-dimensional remains, allowing for its proportions to be accurately reconstructed (McGowan and Motani 2003). It is typically found in shallow marine deposits where the water was probably hundreds of meters deep, so its inclusion in a coastal scene must be regarded as speculative. However, this idea is not unreasonable given that many open-water species routinely visit coastlines to find food or sheltered areas to reproduce. It also allowed the scene to contain some rocky shore elements not commonly associated with marine reptile art—a useful means to avoid repetition in a book with a number of marine scenes.

ANCHIORNIS: A DINOSAUR THAT WAS ALMOST A BIRD (JURASSIC)

Although *Anchiornis* was discovered in only the last few years, its life appearance is already far better understood than that of most other fossil species. *Anchiornis* is known from exceptional fossils that record minute details of feather structure and distribution (e.g., Saitta, Gelernter, and Vinther 2017), and it is also the subject of several studies reconstructing its color pattern using fossil pigment cells (Li et al. 2010; Lindgren et al. 2015). These fossil pigments allow us to be confident that *Anchiornis* probably looked something like the reconstruction here: generally black and white, with a mottled appearance on its inner wing, a stripy distal wing, and some ruddy-colored feathers on its head. It's important to emphasize that not all pigments preserve in fossils, and it remains to be seen whether the blacks, whites, and reds of *Anchiornis* were filtered through other mechanisms of color production; but we nevertheless have a strong foundation for the life appearance of this animal.

BRONTOSAURUS (JURASSIC)

Our numbers of named sauropod species have exploded in recent decades, and it may seem strangely conservative to include an old-school taxon like *Brontosaurus* in a modern paleoart project. But *Brontosaurus* has a taxonomic tie to *LTTA*, as well as a completely contrasting modern interpretation to Knight's era of dinosaur science, making it an ideal subject for this book. Paul (2016) and Scott Hartman's Skeletal Drawing website give excellent modern takes on sauropod form. The idea of rearing, neck-wrestling sauropods would have surely seemed outlandish—maybe ridiculous—to early twentieth-century scholars, who saw sauropods as sluggish, swampbound creatures, yet today the idea is entirely in keeping with what we know of *Brontosaurus* anatomy and biomechanics. See Taylor et al. (2015) for more about *Brontosaurus* neck combat.

MESOZOIC MAMMALS (CRETACEOUS)

This scene of Early Cretaceous Britain features fauna of the Purbeck Group. The mammals from this unit are known only from teeth, so their form is heavily inspired by better-known relatives from China, mostly *Juramaia sinensis* (Luo et al. 2011). The predatory dromaeosaur in the center of the image—*Nuthetes destructor*—is also known from scant material: teeth in a lower jaw. Sauropods are represented from the same deposits as these animals through their footprints (Sweetman, Smith, and Martill 2017). Their arrangement in this scene is a nod to Knight's *LTTA Brontosaurus* illustration.

YUTYRANNUS, THE FEATHERED TYRANT (CRETACEOUS)

The composition of this picture is a direct transplant from Knight's *LTTA Tyrannosaurus* image, where one tyrant looms over a smaller competitor. A creative solution was needed in order to elevate the head of the foremost *Yutyrannus* on account of proportional differences between *Yutyrannus* and *Tyrannosaurus*, as well as the slightly over-long tyrannosaur body in Knight's original painting. An energetic leap from the foremost animal solved this problem and captured our modern view of dinosaur behavior and physiology—depictions of active dinosaurs were rare in Knight's time but are *de rigueur* today. It's become something of a trope to show *Yutyrannus* in a snowscape nowadays, and it did not live in a perpetually snowy landscape, but few compositions sell our new age in dinosaur paleontology better than fluffy tyrannosauroids romping around in snow and ice. The anatomy of *Yutyrannus* is well understood—see Xu et al. (2012) for details.

FLOWERS AND INSECT POLLINATION (CRETACEOUS)

This scene depicts the flora of the Brazilian Crato Formation, a site of exceptional preservation that also has a tremendous wealth of fossil insects. The principle plant in the scene is *Araripia florifera*, an early angiosperm, and was based on descriptive work by Mohr and Eklund (2003). The insects—cockroaches, damselflies, and scoliid wasps—are based on crushed but exceptionally preserved Crato insects, all of which have close modern relatives to assist their reconstruction. The extensive documentation of the Crato biota by Martill, Bechly, and Loveridge (2007) provided references for the insects in this scene.

CRETOXYRHINA AND PTERANODON (CRETACEOUS)

Pteranodon is one of the best-represented flying reptiles, with well over eleven hundred specimens known, so its basic form can be reconstructed with confidence (Bennett 2001). More surprising is the fact that the skeleton of *Cretoxyrhina* is also known. The cartilage skeletons of sharks rarely preserve, but we have several excellent skeletons of *Cretoxyrhina* that record body length, fin size, and skull proportions (Shimada 1997). As noted in the main text, this image is based on a fossilized interaction between these shark and pterosaur species: a *Pteranodon* neck vertebra found in tight association with a tooth from a midsized *Cretoxyrhina*

(see Hone, Witton, and Habib 2018). It is impossible to know whether this association resulted from scavenging or predation: with no data to guide us either way, the rule of cool won out in this depiction of dramatic breaching predation.

MIGHTY ZUUL, DESTROYER OF SHINS (CRETACEOUS)

The ankylosaur *Zuul crurivastator* is named after a doglike creature from a film that science has conclusively established is one of the most enjoyable of all time: *Ghostbusters* (1984). *Zuul* is a relatively well-known animal with a terrific three-dimensional skull; a well-preserved tail; and numerous skin details, which include scales and sheaths (Arbour and Evans 2017). My illustration may become dated as more of the *Zuul* specimen is prepared, since much of the body has been recovered but is still, at time of writing, being removed from rock. *Ghostbusters* fans may notice a few very subtle homages to the movie in this scene.

GIANT SEA LIZARDS AND THE MESOZOIC MARINE REVOLUTION (CRETACEOUS)

A longstanding convention of nineteenth- and early twentieth-century paleoart was to show marine animals swimming through surface waters, irrespective of their ability for submerged locomotion. Knight's *LTTA* mosasaur was no exception, and the *Globidens* in this image are a tribute to this now-unfashionable portrayal of these animals. Their soft tissues have been closely based on monitor lizards, as might be particularly obvious around their faces, and also reference exceptionally preserved mosasaur specimens that show details of their body outlines and skin (e.g., Lindgren et al. 2010; Lindgren, Kaddumi, and Polcyn 2013).

A COLOSSAL AMMONITE (CRETACEOUS)

As alluded to in the main text, our knowledge of ammonite soft-tissue anatomy is extremely limited, and the reconstruction shown here is a fairly conventional artistic take on the ammonite form. Possible soft-tissue remains of Cretaceous ammonites from Germany hint at a low number of short tentacles, which I've followed here (ten tentacles in total), and I have assumed complex eyes akin to those of squid and octopus rather than the simple eyes of *Nautilus* (see Klug and Lehman [2015] for a review of what is known of ammonoid soft tissues). *Parapuzosia* probably lived for some time to reach a shell diameter of 2–3 m, and I've speculated that their shells were colonized by algae and other encrusting organisms, as we see occurring in long-lived whales and turtles today.

GIANT FLYING REPTILES (CRETACEOUS)

The sorry state of giant azhdarchid pterosaur fossils is alluded to in the main text, but it's a point worth repeating here: depending on the species, either we have scraps of bone from different parts of the skeleton, or—in the case of the famous giant *Quetzalcoatlus northropi*—we have a giant, incomplete left wing. Accordingly, any reconstruction you see of these giants is a smaller azhdarchid that has been scaled up to the size of a small airplane. Our limited data suggests that this approach is not entirely inappropriate, and we know something of how pterosaur bones scale to facilitate flight at great size, but we await complete giant azhdarchid fossils to confirm our artistic takes on their overall form. The azhdarchid anatomy shown in this image is based on my work on these animals published with Darren Naish and Mike Habib (e.g., Witton and Naish 2008; Witton and Habib 2010; Naish and Witton 2017).

DEINOSUCHUS, AN ENORMOUS ALLIGATOROID (CRETACEOUS)

Though smaller specimens of *Deinosuchus* are known from relatively complete remains, the largest animals are represented by lesser material, mainly bony scutes and partial skulls. As with so many prehistoric giants, our depictions of the largest individuals are effectively scaled-up smaller specimens. As noted in the main text, the skull of modern *Deinosuchus* reconstructions is somewhat more alligator-like than is seen in older reconstructions, reflecting new finds of more complete skulls. The flipper in the *Deinosuchus* mouth belongs to a large turtle, in line with our understanding of their most common prey. Schwimmer (2002) provides an excellent overview of *Deinosuchus* paleobiology and has informed much of my depiction of this animal.

TRICERATOPS (CRETACEOUS)

As indicated in the main text, skin of this *Triceratops* is a little unusual, but it is based as much as possible on *Triceratops* skin samples held at the Houston Museum of Natural Science. At time of writing, these have yet to be formally described, and some details of this image may warrant revision after details of these specimens are publicly available. The shape of the horns is also atypical, but it reflects horn shape changes recorded in *Triceratops* growth series as well as the mechanics of horn growth in living animals. Horn sheaths grow throughout life, with their tips being old parts of sheath that have been pushed off the horn skeleton by newly deposited horn tissue. In *Triceratops*, juvenile animals have upturned horns, dictating that the sheath tips in adult animals likely bore an upward curve. The result is a rather more elaborate horn shape than we're used to, but one that might be more defensible than traditional horn reconstructions (Witton 2018).

THE CRETACEOUS–PALEOGENE EXTINCTION (CRETACEOUS)

The biggest issue with restoring the events of the end Cretaceous impact event is scale. Though it is one of the few extinction events with illustratable physical phenomena, the impact event dwarfs even the largest organisms to such an extent that drawing a biotic response to it is challenging. The sauropod at the base of this scene is the 30-m-long titanosaur *Alamosaurus sanjuanensis*, and it's still tiny compared to the oncoming wave and the growing thermal plume. If this image doesn't instill a sense of impending mortality in the face of planetary forces, bear in mind that the wave is not illustrated at full size. The impact scenario depicted here was primarily informed by Schulte et al. (2010).

GASTORNIS, A GIANT LAND WATERFOWL (EOCENE)

Though most species of *Gastornis* are relatively fragmentary, the osteology of the American species *Gastornis gigantea* is well known and has been recorded in detail (Matthew, Granger, and Stein 1917). The soft tissue of *Gastornis* is almost entirely unrepresented in fossils, however, except for a giant isolated feather from the Green River Formation, which is probably referable to the genus. If so, it justifies a reconstruction that has neater feathering than the ratite-like shaggy feathers that are typical in older illustrations (Naish 2016). *Gastornis* hatchling material also remains elusive, so they were speculatively restored based on the precocial nature of anseriform offspring, as well as by analogy to the hatchlings of living flightless birds.

ONYCHONYCTERIS, AN EARLY BAT (EOCENE)

Complete fossils of *Onychonycteris* provide excellent proportional detail, but they are crushed dorsoventrally, leaving aspects of the skull poorly known (Simmons et al. 2008). I accordingly based the appearance of the *Onychonycteris* face on other early chiropteran fossils as well as the morphology of living megabats, these seemingly being superior anatomical analogues for Eocene bats than the more specialized, echolocating microbats. The chosen pose reflects the quadrupedal takeoff mechanic used by many modern bats (e.g., Schutt et al. 1997): if *Onychonycteris* spent a good deal of time on the ground, as is suspected, it may have employed this launch strategy often.

ARSINOITHERIUM AND THE EVOLUTION OF GIANT MAMMALS (EOCENE)

Most *Arsinoitherium* species are represented by just a few teeth and jaw bones, but *Arsinoitherium zitteli* is known from essentially complete skeletal remains and multiple skulls that give insights into the appearance of animals at different growth stages. Museum-mounted skeletons and skulls informed much of this image, but details of adult and juvenile skull shapes and textures were taken from Andrews (1906). The elaborate horns of the adult in this picture are larger than suggested by the skull, but they're consistent with the way horn sheaths grow in living animals—the overlying keratin often adds considerable length to the underlying horn skeleton.

GEORGIACETUS, THE LAST LAND WHALE (EOCENE)

Our collection of a smattering of bones from across the *Georgiacetus* skeleton makes it a reasonably restorable animal, although some aspects of its anatomy must be borrowed from other, more completely known protocetids, such as *Maiacetus inuus*. Hulbert (1998) and Uhen (2008) were principle references for my illustrations, with other details taken from Marx, Lambert, and Uhen (2016). The current absence of *Georgiacetus* calf fossils meant the juvenile in this scene was based on the expected proportions and demeanor of cetacean calves, both extant and fossil (Gingerich et al. 2009). Both have substantial amounts of body fat to smooth out their contours. The frigate bird snuck into the bottom of the scene references fossils of these birds occurring in Eocene rocks from the southern United States.

PARACERATHERIUM, A GIANT RHINOCEROTOID (OLIGOCENE)

Numerous interpretations of *Paraceratherium* anatomy have existed over the last century, leading to reconstructions looking like scaled-up rhinos or robust-looking giraffes. Most recent literature seems to have found a middle ground between these two extremes, with a form resembling a gigantic, somewhat horselike creature. As noted in the main text, there is good reason to assume that *Paraceratherium* bore a short proboscis. Paul's (1997) skeletal was the primary anatomical reference for this painting, with some additional details taken from Prothero (2013).

DAEODON, A FORMIDABLE PIGLIKE CREATURE (MIOCENE)

Daeodon anatomy and proportions are well documented, so we can be confident that their heads really were that large compared to their bodies. A mounted *Daeodon* skeleton in the American Museum of Natural History informed much of this image, with some details about anatomy, skeletal function, and behavior taken from Joeckel (1990) and Sundell (1999). The vultures in this scene are based on recent discoveries in Miocene sediments of North America (e.g., *Anchigyps voorhiesi*), which suggest that Old World (or gypaetine) vultures were present in North America as early as the Miocene, if not the late Oligocene, although their remains are not yet sufficiently known for a precise reconstruction (Zhang, Feduccia, and James 2012). They are devouring the remains of *Menoceras*, a small, twin-horned rhinocerotid.

DEINOTHERIUM, THE CHIN-TUSKER (MIOCENE)

Excellent skeletal remains of *Deinotherium giganteum* allow its basic anatomy to be reconstructed in detail, though (as mentioned in the main text) some controversy persists over the length of its trunk. I have followed the short-trunk interpretation here, as outlined by Markov, Spassov, and Simeonovski (2001). Other artists (e.g., Antón 2003) justify a longer trunk because of the need to reach water to drink, which seems like an equally sensible approach to reconstructing this animal. I have made efforts with this image to give *Deinotherium* a hair and color scheme unlike that of a living elephant, given that—though on the same evolutionary line—they are not especially closely related.

THE KILLER SPERM WHALE *LIVYATAN* (MIOCENE)

Neither animal in this scene is especially well represented by fossils. *Livyatan* is known only from skull material and teeth, leaving much about its body shape and size mysterious (Lambert et al. 2010). I have based its appearance here on pygmy and dwarf sperm whales, as they seem more representative of early sperm whale form than the true sperm whale, which has many derived and unusual features. As noted in the text, the anatomy of *Otodus megalodon* is extremely poorly known, and even reconstructing its decaying jaws requires some degree of imagination.

THE AQUATIC SLOTH, *THALASSOCNUS* (MIOCENE–PLIOCENE)

Although *Thalassocnus* is a relatively recent discovery, enough has been learned about its osteology to permit reconstructions of its skeleton. My reconstruction has been informed most by a mounted skeleton in Muséum national d'histoire naturelle, Paris, with facial anatomy based on the work of Bargo, Toledo, and Vizcaíno (2006). It is not clear how the soft-tissue anatomy of *Thalassocnus* was modified to aquatic life. Modern semiaquatic species that could be considered analogues to *Thalassocnus* are varied in this regard, particularly concerning the presence and absence of structures like hair. I restored *Thalassocnus* with hair because many semiaquatic animals—bears, capybaras, beavers, otters—retain it, though a reconstruction with less hair would also be defensible because many aquatically adapted mammals—seals, cetaceans, hippos, and so on—have reduced furry coats. The abundance of kelp in this scene reflects the appearance of kelp forests during the Miocene.

GRASSES, NEAR-HYENAS, AND HORSES (MIOCENE)

Reconstructions of both mammal species in this scene are aided by well-understood osteologies of both *Dinocrocuta* and *Hipparion*. My *Dinocrocuta* was strongly influenced by anatomical reconstructions of this taxon performed by Mauricio Antón (2016). This scene is easily the most visceral illustration in this book, and it is an attempt to depict a more realistic predator–prey interaction than the gladiatorial slugging matches that we often see in art of prehistoric subjects. The manner in which *Hipparion* has been disemboweled is based on the result of chase-and-bite tactics employed by modern hyenas and large canids. These species tend to target the flanks and hind-limb muscles to immobilize prey, rather than pinning and killing it with strategically placed bites to the head or neck.

PELAGORNIS, THE LARGEST FLYING BIRD (MIOCENE)

Unlike many giant extinct fliers, our understanding of *Pelagornis* skeletal anatomy is relatively complete, allowing for confidence in its basic proportions (e.g., Mayr and Rubilar-Rogers 2010; Ksepka 2014). We have no insight from fossils into its reproductive behavior, however, and uncertainty about its relationships to living birds means we cannot make reliable predictions about this aspect of its life based on living relatives. I persevered with this speculative nesting scene because so much pelagornithid art shows them engaging in relatively unassuming behavior, such as soaring or floating on water, while depicting these animals in other roles—such as being parents—is somewhat more interesting and thought-provoking.

GIGANTOPITHECUS (PLEISTOCENE)

As mentioned in the main text, *Gigantopithecus* anatomy is extremely poorly represented by fossils, and any reconstruction—including my own—is extremely speculative. My take is little more than an exercise in rendering a particularly large, quadrupedal, herbivorous ape with some pongine features, and I'm happy to concede that the actual animal may have looked radically different. The other primate in this image, *Homo erectus*, is more robustly represented by fossils, including good examples of its technology.

INSECT SOCIETIES AND THE GIANT ASIAN PANGOLIN (PLEISTOCENE)

The giant Asian pangolin is not a well-known fossil animal, but its anatomical similarity to extant *Manis* species provides artists with some excellent living reference material. Though not as closely related, the extant giant African pangolin (*Smutsia gigantea*) is an especially useful reference taxon for understanding how pangolin anatomy scales up to body lengths of 2 m or more. The *Macrotermes* mound in this scene is also based on extant examples over fossil remains, the surface structures of these mounds rarely fossilizing in a way useful for paleoartistry.

WOOLLY MAMMOTHS (PLEISTOCENE)

This image is another reworking of a *LTTA* original. Knight's take is a more aggressive scene, where Neanderthals are attacking a mammoth that is dynamically turning to face them. My modern version has attempted to show Neanderthals as more sophisticated animals that did not exist purely to hunt big game, but also had curious and protective instincts. Thanks to numerous frozen carcasses preserving mammoth tissues down to their blood cells, woolly mammoths are among the most restorable of all extinct animals: hair length, color, ear shape, trunk length and morphology, the presence of shoulder and head humps, and other features can be translated directly from fossils. Many of these details are now textbook knowledge, but special mention should go to Tridico et al. (2014) for their work showing that mammoth hair was mottled light browns, yellows, and other pale hues, and not the uniform brown or red of countless palaeoartworks (the reds and browns of fossil mammoth hair are actually a result of decay and not indicative of the *in vivo* colors created by their pigment cells). Neanderthal skeletons, technology, art, and genetics are well understood, allowing for reasonable restoration of their guise. The red hair and pale skin of the Neanderthal family in this illustration are based on genetic data showing that the "classic" visage of Neanderthals as dark-skinned, dark-haired humans—an appearance popularized by Knight—is erroneous (Lalueza-Fox et al. 2007). Their clothing is speculative, but it is depicted in keeping with their advanced culture and technology.

A DWARFED GIANT HORNED TURTLE (PLEISTOCENE)

Complete and articulated *Meiolania platyceps* skeletons are on display in several museums, and its anatomy is entirely documented in extensive monographs (e.g., Gaffney 1996). This, and the fact that turtle anatomy records scale impressions, allows some confidence about the basic life appearance of this animal. It is not known how *Meiolania* nested, but laying eggs in a simple scrape in the ground and then covering them in loose sediment seems a reasonable inference based on the behavior of modern turtles.

MOA (HOLOCENE)

Because moas went extinct only recently, their remains are relatively abundant and of high quality. This allows reconstructions of *Dinornis robustus* to be fairly detailed, from the distribution of their feathers to the color of their plumage. Artists also benefit from some areas of New Zealand's moa habitat remaining largely unchanged in the last few hundred years. Many sources informed this image, including skeletal mounts, monographic descriptions of moa soft tissues (e.g., Rawlence et al. 2013), and overviews of moa biology (Angst and Buffetaut 2017).

MODERN HUMANS (HOLOCENE)

No other image in this book proved more of a compositional challenge than our own scene. I knew that I did not want to follow tradition and show humanity as cave-living artists or as conquerors of nature, as such depictions do little to stress than our modern world—with our iPads, skyscrapers, and reality TV shows—is still part of the story of evolution on Earth. I decided to focus this image not on ourselves, but on the aspect of our lives that is likely our longest-lasting legacy to the planet: our waste. Some may consider this a bleak, cynical, or depressing note on which to end our story, and I can't disagree with that assessment. But I hope it's also thought-provoking and a cause for reflection—and an appropriate tonal contrast with the more optimistic nene illustration that closes the narrative.

Literature Cited

Andrews, C. W. (1906). A descriptive catalogue of the Tertiary Vertebrata of the Fayûm, Egypt. *Publications of the British Museum of Natural History.* London, England: Trustees of the British Museum.

Angst, D., & Buffetaut, E. (2017). *Palaeobiology of giant flightless birds.* London, England: iSTE Press/Elsevier.

Antón, M. (2003). Reconstructing fossil mammals: Strengths and limitations of a methodology. *Palaeontological Association Newsletter, 53,* 55–65.

Antón, M. (2016, January 12). Sabertooth's bane: Introducing *Dinocrocuta. Chasing Sabertooths* [weblog message]. Retrieved from https://chasing sabretooths.wordpress.com/2016/01/12/sabertooths-bane-introducing -dinocrocuta/

Arbour, V. M., & Evans, D. C. (2017). A new ankylosaurine dinosaur from the Judith River Formation of Montana, USA, based on an exceptional skeleton with soft tissue preservation. *Royal Society Open Science, 4*(5), 161086.

Bargo, M. S., Toledo, N., & Vizcaíno, S. F. (2006). Muzzle of South American Pleistocene ground sloths (Xenarthra, Tardigrada). *Journal of Morphology, 267*(2), 248–263.

Bennett, S. C. (2001). The osteology and functional morphology of the Late Cretaceous pterosaur *Pteranodon* Part I. General description of osteology. *Palaeontographica Abteilung A, 260,* 1–112.

Berman, J. C. (2003). A note on the paintings of prehistoric ancestors by Charles R. Knight. *American anthropologist, 105*(1), 143–146.

Boyle, J., & Ryan, M. J. (2017). New information on *Titanichthys* (Placodermi, Arthrodira) from the Cleveland Shale (Upper Devonian) of Ohio, USA. *Journal of Paleontology, 91*(2), 318–336.

Brown, B. (1941). The methods of Walt Disney productions. *Transactions of the New York Academy of Sciences,* 3(Series II), 100–105.

Carroll, R. L. (2009). *The rise of amphibians: 365 million years of evolution.* Baltimore, MD: Johns Hopkins University Press.

Cheng, L., Chen, X.-H., Shang, Q. H., & Wu, X.-C. (2014). A new marine reptile from the Triassic of China, with a highly specialized feeding adaptation. *Naturwissenschaften, 101*(3), 251–259.

Chun, L., Rieppel, O., Long, C., & Fraser, N. C. (2016). The earliest herbivorous marine reptile and its remarkable jaw apparatus. *Science Advances, 2*(5), e1501659.

Clark, C. A. (2008). *God—or gorilla: Images of evolution in the Jazz Age.* Baltimore, MD: Johns Hopkins University Press.

Corbacho, J., & Vela, J. A. (2010). Giant trilobites from lower Ordovician of Morocco. *Batalleria, 15,* 3–34.

Czerkas, S. M., & Glut, D. F. (1982). *Dinosaurs, mammoths, and cavemen: The art of Charles R. Knight.* New York, NY: E. P. Dutton.

Davidson, J. P. (2008). *A history of paleontology illustration.* Bloomington, IN: Indiana University Press.

DeBraga, M. (2003). The postcranial skeleton, phylogenetic position, and probable lifestyle of the Early Triassic reptile *Procolophon trigoniceps. Canadian Journal of Earth Sciences, 40*(4), 527–556.

Erwin, D. H., & Valentine, J. W. (2013). *The Cambrian explosion: The construction of animal biodiversity.* W. H. Freeman.

Ferrón, H. G., Martínez-Pérez, C., & Botella, H. (2017). Ecomorphological inferences in early vertebrates: Reconstructing *Dunkleosteus terrelli* (Arthrodira, Placodermi) caudal fin from palaeoecological data. *PeerJ, 5,* e4081.

Gaffney, E. S. (1996). The postcranial morphology of *Meiolania platyceps* and a review of the Meiolaniidae. *Bulletin of the American Museum of Natural History, 229,* 1–116.

Gingerich, P. D., ul-Haq, M., von Koenigswald, W., Sanders, W. J., Smith, B. H., & Zalmout, I. S. (2009). New protocetid whale from the middle Eocene of Pakistan: Birth on land, precocial development, and sexual dimorphism. *PLoS One, 4*(2), e4366.

Gower, D. J., Hancox, P. J., Botha-Brink, J., Sennikov, A. G., & Butler, R. J. (2014). A new species of *Garjainia* Ochev, 1958 (Diapsida: Archosauriformes: Erythrosuchidae) from the Early Triassic of South Africa. *PLoS One*, 9(11), e111154.

Harryhausen, R., Dalton, T., & Bradbury, R. (2004). *Ray Harryhausen: An animated life*. New York, NY: Billboard Books.

Hartman, S. (2016). Taking a 21st century look at *Dimetrodon*. Scott Hartman's Skeletal Drawing.com. Retrieved from http://www.skeletaldrawing.com/home/21stcenturydimetrodon

Hulbert, R. C. (1998). Postcranial osteology of the North American middle Eocene protocetid *Georgiacetus*. In J. G. M. Thewissen (Ed.), *The emergence of whales* (pp. 235–267). Boston, MA: Springer.

Hone, D. W., Witton, M. P., & Habib, M. B. (2018). Evidence for the Cretaceous shark *Cretoxyrhina* mantelli feeding on the pterosaur *Pteranodon* from the Niobrara Formation. *PeerJ*, 6, e6031.

Ivantsov, A. Y. (2009). New reconstruction of *Kimberella*, problematic Vendian metazoan. *Paleontological Journal*, 43, 601–611.

Jenkins Jr., F. A., & Parrington, F. R. (1976). The postcranial skeletons of the Triassic mammals *Eozostrodon*, *Megazostrodon* and *Erythrotherium*. *Philosophical Transactions of the Royal Society B: Biological Sciences*, 273(926), 387–431.

Jeram, A. J. (1993). Scorpions from the Viséan of East Kirkton, West Lothian, Scotland, with a revision of the infraorder Mesoscorpionina. *Earth and Environmental Science Transactions of The Royal Society of Edinburgh*, 84(3-4), 283–299.

Jiang, D. Y., Motani, R., Huang, J. D., Tintori, A., Hu, Y. C., Rieppel, O., . . . & Zhang, R. (2016). A large aberrant stem ichthyosauriform indicating early rise and demise of ichthyosauromorphs in the wake of the end-Permian extinction. *Scientific Reports*, 6, 26232.

Joeckel, R. M. (1990). A functional interpretation of the masticatory system and paleoecology of entelodonts. *Paleobiology*, 16(4), 459–482.

Keppler, M., Benisty, M., Müller, A., Henning, T., van Boekel, R., Cantalloube, F., . . . & Weber, L. (2018). Discovery of a planetary-mass companion within the gap of the transition disk around PDS 70. *Astronomy & Astrophysics*, 617, A44.

Klug, C., & Lehmann, J. (2015). Soft part anatomy of ammonoids: Reconstructing the animal based on exceptionally preserved specimens and actualistic comparisons. In C. Klug, D. Korn, K. De Baets, I. Kruta, & R. H Mapes (Eds.), *Ammonoid Paleobiology: From anatomy to ecology* (pp. 507–529). Dordrecht, Netherlands: Springer.

Knight, C. R. (1935). *Before the dawn of history*. New York, NY: Whittlesey House, McGraw-Hill.

Knight, C. R. (1946). *Life through the ages*. New York, NY: Alfred A. Knopf.

Knight, C. R. (2001). *Life through the ages: A commemorative edition*. Bloomington, IN: Indiana University Press.

Knight, C. R. (1947). *Animal anatomy and psychology for artists and laymen*. New York, NY: Whittlesey House, McGraw-Hill.

Knight, C. R. (1949). *Prehistoric man, the great adventurer: The saga of man's beginnings in word and picture*. New York, NY: Appleton-Century-Crofts.

Knight, C. R. (1959). *Animal drawing: Anatomy and action for artists*. New York, NY: Dover Publications.

Knight, C. R. (2005). *Charles R. Knight: Autobiography of an artist*. Ann Arbor, MI: G.T. Labs.

Ksepka, D. T. (2014). Flight performance of the largest volant bird. *Proceedings of the National Academy of Sciences*, 111(29), 10624–10629.

Lalueza-Fox, C., Römpler, H., Caramelli, D., Stäubert, C., Catalano, G., Hughes, D., . . . & Hofreiter, M. (2007). A melanocortin 1 receptor allele suggests varying pigmentation among Neanderthals. *Science*, 318(5855), 1453–1455.

Lambert, O., Bianucci, G., Post, K., de Muizon, C., Salas-Gismondi, R., Urbina, M., & Reumer, J. (2010). The giant bite of a new raptorial sperm whale from the Miocene epoch of Peru. *Nature*, 466, 105–108.

Lautenschlager, S., Gill, P., Luo, Z. X., Fagan, M. J., & Rayfield, E. J. (2017). Morphological evolution of the mammalian jaw adductor complex. *Biological Reviews of the Cambridge Philosophical Society*, 92(4), 1910–1940.

Lescaze, Z., & Ford, W. (2017). *Paleoart: Visions of the prehistoric past*. Cologne, Germany: Taschen.

Li, Q., Gao, K. Q., Vinther, J., Shawkey, M. D., Clarke, J. A., D'Alba, L., . . . & Prum, R. O. (2010). Plumage color patterns of an extinct dinosaur. *Science*, 327(5971), 1369–1372.

Lindgren, J., Caldwell, M. W., Konishi, T., & Chiappe, L. M. (2010). Convergent evolution in aquatic tetrapods: Insights from an exceptional fossil mosasaur. *PloS One*, 5(8), e11998.

Lindgren, J., Kaddumi, H. F., & Polcyn, M. J. (2013). Soft tissue preservation in a fossil marine lizard with a bilobed tail fin. *Nature Communications*, 4, no. 2423.

Lindgren, J., Sjövall, P., Carney, R. M., Cincotta, A., Uvdal, P., Hutcheson, S. W., . . . & Godefroit, P. (2015). Molecular composition and ultrastructure of Jurassic paravian feathers. *Scientific Reports*, 5, 13520.9-5.

Long, J. A. (2010). *The rise of fishes: 500 million years of evolution* (2nd ed.). Baltimore, MD: Johns Hopkins University Press.

Luo, Z. X., Yuan, C.-X., Meng, Q.-J., & Ji, Q. (2011). A Jurassic eutherian mammal and divergence of marsupials and placentals. *Nature*, 476, 442–445.

Markov, G. N., Spassov, N., & Simeonovski, V. (2001). A reconstruction of the facial morphology and feeding behaviour of the deinotheres. In G. Cavarretta, P. Gioia, M. Mussi, & M. R. Palombo (Eds.), *The world of elephants: Proceedings of the First International Congress, Rome* (pp. 652–655). Rome, Italy: Consiglio Nazionale delle Ricerche.

Märss, T., Turner, S., & Karatajūtė-Talimaa, V. (2007). "Agnatha" II: Thelodonti. In H.-P. Schultze & O. Kuhn (Eds.), *Handbook of Paleoichthyology* (Vol. 1B). Munich, Germany: Verlag Dr Friedrich Pfeil.

Martill, D. M., Bechly, G., & Loveridge, R. F. (Eds.). (2007). *The Crato fossil beds of Brazil: Window into an ancient world*. Cambridge, England: Cambridge University Press.

Marx, F. G., Lambert, O., & Uhen, M. D. (2016). *Cetacean paleobiology*. Oxford, England: Wiley.

Matthew, W. D., Granger, W., & Stein, W. (1917). The skeleton of *Diatryma*, a gigantic bird from the Lower Eocene of Wyoming. *Bulletin of the American Museum of Natural History, 37*, 307–326.

Mayr, G., & Rubilar-Rogers, D. (2010). Osteology of a new giant bony-toothed bird from the Miocene of Chile, with a revision of the taxonomy of Neogene Pelagornithidae. *Journal of Vertebrate Paleontology, 30*(5), 1313–1330.

McGowan, C., & Motani, R. (2003). Ichthyopterygia. In H. D. Sues (Ed.), *Handbook of Paleoherpetology* (Part 8). Munich, Germany: Verlag Dr. Friedrich Pfeil.

Milner, R. (2012). *Charles R. Knight: The artist who saw through time*. New York, NY: Abrams.

Mohr, B. A., & Eklund, H. (2003). *Araripia florifera*, a magnoliid angiosperm from the Lower Cretaceous Crato Formation (Brazil). *Review of Palaeobotany and Palynology, 126*(3–4), 279–292.

Motani, R., Jiang, D. Y., Chen, G. B., Tintori, A., Rieppel, O., Ji, C., & Huang, J. D. (2015). A basal ichthyosauriform with a short snout from the Lower Triassic of China. *Nature, 517*, 485–488.

Müller, A., Keppler, M., Henning, T., Samland, M., Chauvin, G., Beust, H., . . . & Zurlo, A. (2018). Orbital and atmospheric characterization of the planet within the gap of the PDS 70 transition disk. *Astronomy & Astrophysics, 617*, no. L2.

Naish, D. (2016). Lance Grande's *The lost world of Fossil Lake*. *Scentific American* Tetrapod Zoology [weblog message]. Retrieved from https://blogs.scientificamerican.com/tetrapod-zoology/lance-grande-s-the-lost-world-of-fossil-lake

Naish, D., & Witton, M. P. (2017). Neck biomechanics indicate that giant Transylvanian azhdarchid pterosaurs were short-necked arch predators. *PeerJ, 5*, e2908.

Nasterlack, T., Canoville, A., & Chinsamy, A. (2012). New insights into the biology of the Permian genus *Cistecephalus* (Therapsida, Dicynodontia). *Journal of Vertebrate Paleontology, 32*(6), 1396–1410.

Paul, G. S. (1996). The art of Charles R. Knight. *Scientific American, 274*(6), 86–93.

Paul, G. S. (1997). Dinosaur models: The good, the bad, and using them to estimate the mass of dinosaurs. In D. L. Wolberg, E. Stump, & G. D. Rosenberg (Eds.), *DinoFest International Proceedings* (pp. 129–154). Philadelphia, PA: Academy of Natural Sciences.

Paul, G. S. (2016). *The Princeton field guide to dinosaurs* (2nd ed.). Princeton, NJ: Princeton University Press.

Pinheiro, F. L., França, M. A., Lacerda, M. B., Butler, R. J., & Schultz, C. L. (2016). An exceptional fossil skull from South America and the origins of the archosauriform radiation. *Scientific Reports, 6*, 22817.

Prothero, D. R. (2013). *Rhinoceros giants: The paleobiology of Indricotheres*. Bloomington, IN: Indiana University Press.

Rawlence, N. J., Wood, J. R., Scofield, R. P., Fraser, C., & Tennyson, A. J. (2013). Soft-tissue specimens from pre-European extinct birds of New Zealand. *Journal of the Royal Society of New Zealand, 43*(3), 154–181.

Regal, B. (2002). *Henry Fairfield Osborn: Race and the search for the origins of man*. Oxfordshire, England: Routledge.

Saitta, E. T., Gelernter, R., & Vinther, J. (2017). Additional information on the primitive contour and wing feathering of paravian dinosaurs. *Palaeontology, 61*, 273–288.

Schulte, P., Alegret, L., Arenillas, I., Arz, J. A., Barton, P. J., Bown, P. R., . . . & Willumsen, P.S. (2010). The Chicxulub asteroid impact and mass extinction at the Cretaceous-Paleogene boundary. *Science, 327*(5970), 1214–1218.

Schutt, W. A., Altenbach, J. S., Chang, Y. H., Cullinane, D. M., Hermanson, J. W., Muradali, F., & Bertram, J. E. (1997). The dynamics of flight-initiating jumps in the common vampire bat *Desmodus rotundus*. *Journal of Experimental Biology, 200*(Pt. 23), 3003–3012.

Schwimmer, D. R. (2002). *King of the crocodylians: The paleobiology of* Deinosuchus. Bloomington, IN: Indiana University Press.

Sennikov, A. G. (2008). Archosauromorpha. In M. F. Ivakhnenko & E. N. Kurochkin (Eds.). *Fossil vertebrates of Russia and adjacent countries: Fossil reptiles and birds Part 1* (pp 266–318). Moscow, Russia: Russian Academy of Sciences Paleontological Institute.

Shimada, K. (1997). Skeletal anatomy of the Late Cretaceous lamniform shark, *Cretoxyrhina mantelli* from the Niobrara Chalk in Kansas. *Journal of Vertebrate Paleontology, 17*(4), 642–652.

Simmons, N. B., Seymour, K. L., Habersetzer, J., & Gunnell, G. F. (2008). Primitive Early Eocene bat from Wyoming and the evolution of flight and echolocation. *Nature, 451*, 818–821.

Sommer, M. (2016). *History within: The science, culture, and politics of bones, organisms, and molecules*. Chicago, IL: University of Chicago Press.

Stout, W. (2002). *Charles R. Knight sketchbook* (Vol. 1). Cambridge, MA: Terra Nova Press.

Stovall, J. W., Price, L. I., & Romer, A. S. (1966). The postcranial skeleton of the giant Permian pelycosaur *Cotylorhynchus romeri*. *Bulletin of the Museum of Comparative Zoology, 135*(1), 1–30.

Sundell, K. A. (1999). Taphonomy of a multiple *Poebrotherium* kill site—An *Archaeotherium* meat cache. *Journal of Vertebrate Paleontology, 19*(3), 79A.

Sweetman, S. C., Smith, G., & Martill, D. M. (2017). Highly derived eutherian mammals from the earliest Cretaceous of southern Britain. *Acta Palaeontologica Polonica, 62*(4), 657–665.

Tapanila, L., Pruitt, J., Pradel, A., Wilga, C. D., Ramsay, J. B., Schlader, R., & Didier, D. A. (2013). Jaws for a spiral-tooth whorl: CT images reveal novel adaptation and phylogeny in fossil *Helicoprion*. *Biology Letters, 9*(2), 20130057.

Taylor, M. P., Wedel, M. J., Naish, D., & Engh, B. (2015). Were the necks of *Apatosaurus* and *Brontosaurus* adapted for combat? *PeerJ PrePrints, 3*, e1663.

Tridico, S. R., Rigby, P., Kirkbride, K. P., Haile, J., & Bunce, M. (2014). Megafaunal split ends: Microscopical characterisation of hair structure and function in extinct woolly mammoth and woolly rhino. *Quaternary Science Reviews, 83*, 68–75.

Uhen, M. D. (2008). New protocetid whales from Alabama and Mississippi, and a new cetacean clade, Pelagiceti. *Journal of Vertebrate Paleontology, 28*(3), 589–593.

Wendruff, A. J., & Wilson, M. V. (2012). A fork-tailed coelacanth, *Rebellatrix divaricerca*, gen. et sp. nov. (Actinistia, Rebellatricidae, fam. nov.), from the Lower Triassic of Western Canada. *Journal of Vertebrate Paleontology, 32*(3), 499–511.

Witton, M. P. (2018). *The palaeoartist's handbook: Recreating prehistoric animals in art*. Marlborough, Wiltshire, England: Crowood Press.

Witton, M. P., & Habib, M. B. (2010). On the size and flight diversity of giant pterosaurs, the use of birds as pterosaur analogues and comments on pterosaur flightlessness. *PloS One, 5*(11), e13982.

Witton, M. P., & Naish, D. (2008). A reappraisal of azhdarchid pterosaur functional morphology and paleoecology. *PLoS One, 3*(5), e2271.

Xu, X., Wang, K., Zhang, K., Ma, Q., Xing, L., Sullivan, C., . . . & Wang, S. (2012). A gigantic feathered dinosaur from the Lower Cretaceous of China. *Nature, 484*, 92–95.

Zhang, Z., Feduccia, A., & James, H. F. (2012). A late Miocene accipitrid (Aves: Accipitriformes) from Nebraska and its implications for the divergence of Old World vultures. *PloS One, 7*(11), e48842.

MARK P. WITTON is a vertebrate paleontologist, a technical consultant on paleontological documentaries, and also a paleoartist, graphic designer, and author. His books include *The Palaeoartist's Handbook: Recreating Prehistoric Animals in Art* and *Pterosaurs: Natural History, Evolution, Anatomy.* He lives in Portsmouth, UK, with nine tetrapods: two lizards, one snake, four chickens, one dog, and one long-suffering, infinitely patient wife.